The Creating BRAIN

The Creating BRAIN

THE NEUROSCIENCE OF GENIUS

Nancy C. Andreasen, M.D., Ph.D.

DANA PRESS New York • Washington, D.C.

Published by Dana Press
New York/Washington, D.C.

The Dana Foundation
745 Fifth Avenue, Suite 900
New York, NY 10151

900 15th Street NW
Washington, DC 20005

ISBN: (cloth) 1-932594-07-8

Library of Congress Cataloging-in-Publication Data
Andreasen, Nancy C. The creating brain : the neuroscience of genius /
Nancy C. Andreasen. p. cm.
Includes bibliographical references and index.
ISBN 1-932594-07-8 (alk. paper)
1. Neuropsychology. 2. Genius. 3. Creative ability. I. Title.
QP398.A53 2005
612.8—dc22
2005009994

Cover Design by FTM Design Studio
Text Design by Kachergis Book Design
Printed by Edwards Brothers Incorporated

www.dana.org

To the "lost geniuses" of the past, in the hope that this book will help more thrive in the future.

CONTENTS

LIST OF ILLUSTRATIONS

PREFACE

When I was a kindergartener, I was IQ-tested and declared a "genius." My parents were not supposed to disclose this diagnosis to me, but of course they did. That was the beginning of my life of yin and yang with respect to having a "gift," starting with my parents who were really of two minds about it.

On the one hand, they were proud of their precocious little daughter. On the other hand, they were also socially conservative about the role of women. They wanted me to excel in school and be the "teacher's pet." But they also thought this should lead to my becoming a kindergarten teacher like my mom, or perhaps a dental hygienist or nurse if I wanted a career in the health sciences. Above all they wanted me to find the right man (probably a doctor), to marry him, and to have a family.

I had other ideas. In fact, as a very little girl, I provided entertainment at my parents' dinner parties by responding to "What do you want to be when you grow up?" with "I want to be the first woman president of the US." That idea was so preposterous that it didn't trouble my parents. But my other ambitions, changing to wiser choices as I grew up, did. I might get a Ph.D. and be a university professor. I might become a fiction writer or a poet. I might go into journalism and work for a major newspaper, perhaps as a foreign correspondent.

My parents despaired over my stubborn refusal to seek a safe and secure profession. My late teens reverberated with their warnings:

"Nancy, you'll never get a job if you just major in history and philosophy and English literature." "Nancy, no one will want to marry you if you get a Ph.D." "Nancy, George is such a nice young man—marry him and settle down." I journeyed on, however, escaping to Harvard on a Woodrow Wilson fellowship and on to Oxford with a Fulbright scholarship, watching my mother fail to hold back her tears as I left home for the first and definitive time. When I decided to go to medical school, it was nearly the end of the world, from the perspective of Mom and Dad. One thing I did do right was to marry George and have two daughters.

Looking back now from the perspective of many years of achievement and success, I am struck by the paradox that my parents valued my intelligence and drive, but not its consequences. Nature and nurture were at odds with one another in my early life. That was my yin and yang.

I have often wondered why I was born with a drive to do something, why I was able to go against the grain and could somehow take risks that my parents perceived to be so dangerous. My parents did eventually grow to be proud of me, as it became clear that I could be both a good mother and a good doctor. Just a few days before he died, my father said to me: "Nancy, we tried our best to hold you back, but we just couldn't do it. You were like a young filly who was determined to win the race. And you did turn out OK after all." I am very glad that he said that, for it lifted away my burden of guilt.

As I was growing up, pulled back and forth by the opposing forces in my life, I wondered what it meant to be a "genius." Of course, I began to realize that I did not deserve that title. Reading Shakespeare, listening to Mozart, or looking at the works of Michelangelo, I could see what genuine genius really is. As I continued to mature, becoming a psychiatrist and a neuroscientist, I also wondered how their brains could produce such magnificent and original achievements. Although Mozart was trained to be a prodigy, so many other great geniuses came from nowhere. How? Why?

I also wondered how many geniuses had been born—had been given the creative nature—but were unable to realize their gifts for lack of nurture. Half of the human beings in history are women, for example, but we have had so few women recognized for their genius. How many were held back by societal influences, similar to the ones I encountered and dared to ignore? I cannot believe that women are innately less creative than men. But the problem goes beyond gender. It includes racism, prejudice, poverty, wars, lack of education, and a host of other forces that prevent the seeds of creativity from sprouting. We cannot afford to waste human gifts. We need to learn how to nurture the creative nature. Therefore, I have chosen to dedicate this book to all those lost geniuses of the past, in the hope that it will help more thrive in the future.

I wish to thank those who helped make this book possible. My administrative assistant Luann Godlove worked patiently and faithfully on the preparation of the manuscript. Jane Nevins of the Dana Press was an invaluable critic and supporter, full of helpful suggestions as she cheered me on to the finish line. And my husband Terry Gwinn was always at my side providing encouragement and support, as "best buddies" should.

The Creating BRAIN

1

THE NATURE OF CREATIVITY

The Ingenious Human Brain

To see a World in a Grain of Sand
And a Heaven in a Wild Flower,
Hold Infinity in the palm of your hand
And Eternity in an hour.

—William Blake, *Auguries of Innocence*

Human beings similar to you and me have lived on our Earth for approximately 100,000 years. If we include the humans who preceded modern man (known these days as *Homo sapiens sapiens*), such as Neanderthal man or *Homo erectus* or even *Homo sapiens,* we have been present even longer. I often wonder what these early people felt, and what their lives were like. These "cavemen" seem far removed from us, partly because they are so often portrayed as slumped and naked and hairy, or wearing rudimentary clothing, and carrying clubs. But they are our ancestors, and *not* all that far removed. And they may have had much more soul and spirit and creativity than the stereotyped images in books and museums lead us to believe.

When did these early people begin to think and feel in the ways that we do in the twenty-first century? When did male and female human beings begin to feel dedicated and faithful love for one another and to mate for life, as many so-called "lower" animals do? When did they begin to mourn the loss of their mate, if he or she died? When did they begin to love their children, as we love ours? How did they feel about growing old? How did they begin to understand their own mortality? When did they begin to bury their dead, and why? When did they begin to wonder about supernatural forces greater than themselves, to develop a sense of spirituality, and to worship gods or a God? When did they begin to map the motions of the stars and planets in the skies and pass this information on to others? When did they begin to tell stories while watching an open fire, to celebrate together with communal feasts, and to create alcoholic beverages to add to the conviviality? When did they begin to wear clothing, and why? As we move mentally back in time, we can ask almost endless questions about what our early ancestors might have been like, and how and when their inner spirits began to resemble our own.

Because we have no written records, we have no answers to these questions. The earliest written records come relatively late in the history of humankind. Cuneiform script and hieroglyphics, the earliest forms of writing, are only about five thousand years old, much more recent than the emergence of modern man.

We do know, even without written records, that some of these prehistoric people possessed the gift of creativity—the capacity to see something new that others could not. Someone picked up a stone and saw a tool. Someone realized that it could be made sharp and pointed by chipping away at it. Someone recognized that a group of people could join together and hunt large food-rich animals, using their collective intellect and strength. Someone suspected that seeds could be planted and crops grown, thereby creating a more secure food supply. Someone figured out how to concentrate light or to chip flints together to create a fire and cook. Someone worked out that circular wheels could facilitate moving heavy objects.

Some of these early creative people must have become storytellers, long before written language existed. They must have begun to remember, retell, and thereby record the history of their tribe or city, passing on their stories and legends through an aural/oral tradition to future generations. Some became artists, as attested by the magnificent drawings in the 17,000 BC Cave of Lascaux in the Dordogne in France, which is presumed to be a type of religious center used to bless the rite of hunting. These artists (or perhaps priests) figured out how to use the colors from the earth or charcoal from a fire, combined with the colors of the stone wall, to create more than one thousand graceful depictions of animal life. Other creative people tracked the movement of the sun and figured out how to place enormous rocks to honor the time of the summer solstice, thereby creating the five-thousand-year-old astronomical and religious monument known as Stonehenge. This required very precise mathematical and astronomical calculations that permitted our ancestors to predict the times of the solar and lunar eclipses. (I was able to visit this amazing monument many years ago when almost no one did, when I was a very young student of English literature at Oxford University and when Stonehenge was a neglected prehistoric site in the middle of an abandoned open field on the Salisbury plain in Wiltshire.)

We have so many amazing examples of human creativity from human prehistory and history, such as the pyramids of Egypt and the Mayan ruins at Chichin Itza, the statues of Nemrut Dag in western Turkey, the Acropolis in Athens with its Parthenon, the Roman roads and aqueducts that crisscross Europe—an ongoing progression of varied and enduring human creative achievements. Music must also have been created at some very early point, and the progression of these creative achievements would almost certainly have been accompanied by diverse kinds of music, the nature of which is forever lost to us.

Even during their early history, human beings have had the spark of creativity. They could see things that did not yet exist. They could imagine. They could yearn for beauty and make it immanent through art and literature and architecture. They could search the skies by day and night to look at the sun, the moon, the planets, and the stars ("the other stars," to be precise), attempting to understand their movements and their mysteries. They could study the animals and plants around them, seeking new ways to use nature to help them survive, and sometimes also seeking ways to record its awe-inspiring beauty. They could create engineering miracles, such as the pyramids, Stonehenge, and the Parthenon. They could imagine beings and forces, greater than themselves, which were guiding and shaping their world. They could even create moral codes that minimized the importance of individual survival and sublimated it to some higher cause. As the English Romantic poet William Blake described it, they knew how . . .

> To see a World in a Grain of Sand
> And a Heaven in a Wild Flower,
> Hold Infinity in the palm of your hand
> And Eternity in an hour.

Where does this creative spark come from? How do fevers in the human brain produce the dreams and visions that become transformed into blazes of insight? Why do our brains yearn after beauty and truth, as if they are all we know on Earth, and all we need to know? Why are some human beings so highly creative that we recognize them as true geniuses? Can those of us who are less creative enhance this innate human creative capacity in ourselves and others? And can we instill it in our children and in future generations?

Those are grand questions. And those grand questions are the subject of this book. I'll be your guide in trying to answer them. I am thrilled that I am able to do it. This book has been bubbling and churning in my brain for nearly thirty years, and I am grateful that I have finally found time to write it. Let me introduce myself to you, so that you will know why I want to be your guide to understanding the fascinating topic of the creating brain and the neuroscience of genius.

Most people think of me as a well-known neuroscientist, or brain scientist. They know me as a person who helped to launch the neuroimaging revolution—the use of brain-imaging technologies to study and measure brain structure and function in living human beings. Neuroimaging is now giving us tools with which we can study many different kinds of human abilities that seemed beyond our reach only a few years ago: how we feel empathy for others, how we can modify the rhythms in our brains when we meditate, or even how we may eventually make greater use of the brain's creative capacities. Understanding how the brain produces our profoundly human abilities to feel and to think is one of my passions as a neuroscientist.

People also know me as a dedicated physician-scientist, a psychiatrist who has devoted much of my life to caring for people with serious mental illnesses and to searching for the mechanisms of these brain diseases, with the hope of improving treatments for them and thereby improving the lives of people who suffer from them. Working on a daily basis with people whose minds are troubled by the intrusion of psychotic thoughts or a concern that they are "losing their minds" teaches a thoughtful psychiatrist a great deal about the nature of the mind and brain. It also raises many questions about how such strange experiences arise in the brain, what triggers them, and how we can minimize them or even turn them off. I have also had a strong interest in the relationship between creativity and mental illness, exploring the age-old speculation that there is a fine line between genius and insanity. I have spent much of my career pondering and studying whether this is true, in what sense it is true, and what this might mean both for the creative person and for society. Because of this interest, I have done one of the few modern studies of creativity and mental illness, which I'll describe later in this book.

So my life in science and as a doctor has given me a rich reservoir of knowledge to share about how our brains think, feel, and create novel ideas and objects.

But I also bring another perspective to this book. I have a deeply rooted passion for the arts that is as great as my passion for science and medicine. I am an M.D. and a Ph.D. My Ph.D. is not in biochemistry or biology, as many assume, but instead in Renaissance English literature. I was an established scholar in the field, and the first book I published was *John Donne: Conservative Revolutionary* (1967). Before deciding to make a career change to medicine and brain science, I spent five years as a professor of English literature at several different universities. In my mind I walk the streets of Elizabethan England or Renaissance Florence as readily as I walk the corridors of hospitals or scientific laboratories. And I read Shakespeare and Sophocles as readily as I read brain scans or *Science* magazine. I also have a great love of the visual arts, of theater, and of dance. This happily diverse background gives me a somewhat unique perspective on creativity, because I have delved deeply into fields that many consider to be almost opposite to one another.

So I'll be at your side as we embark on a wonderful exploratory adventure: examining what the Norwegian playwright Henrik Ibsen once called, "the spark of the divine fire." Our mission is especially exciting and novel because we are not just exploring the nature of creativity, but also the *neuroscience* of creativity, a difficult mission that few have as yet attempted. We are going to be observing how the most interesting and complex organ in the human body, the brain, produces the most fascinating ability that human beings have, the capacity to be creative.

The Evolution of Concepts of Creativity

Any investigation of creativity—whether it be its correlation with brain structure and function, personality, cognitive style, or mental illness—must begin with a fundamentally difficult question. What exactly *is* creativity? How is it related to intelligence?

The word "create" derives from the Latin *creare,* which means "to produce, make, or create." There is an obvious link between a human creator of sonnets or statues and a godly Creator of the Universe. Thus, some of the great creators of the past were honored with the accolade "divine." For example, the Italian biographer of Renaissance artists, Giorgio Vasari, refers to "the divine Michelangelo."

The concept of creativity has historically been used interchangeably with the term "genius." *Genius* is a common Latin word, originally derived from the Greek *ginesthai,* which meant "to be born or come into being." In common usage in Roman times, *genius* originally referred to a god or spirit given to each person at birth that would determine his or her character and fortunes. In early English usage this meaning was sometimes retained, as in Sir Philip Sidney's statement that "a poet, no industry can make, if his own genius be not carried into it."

This meaning soon evolved into the more familiar usage of *genius* as it was employed in later centuries, when people began to use the word to refer to people who had extraordinary intellectual abilities or capacities for imaginative creation. In general, *genius* was the most popular term for referring to great creative capacities during the seventeenth through early twentieth centuries. In the nineteenth century, the Italian psychiatrist Cesare Lombroso wrote a book entitled *Genius and Insanity*, and the English naturalist Francis Galton published *Hereditary Genius*. Early in the twentieth century the English physician Havelock Ellis wrote *A Study of British Genius*. These three books were all efforts to examine the intertwining between heritability of creativity and mental illness. Only with the widespread emergence of the modern discipline of psychology in the twentieth century did people begin to use the term "creativity" and attempt to define it in a scientific way.

Creativity versus Intelligence: "Terman's Geniuses"

The first modern effort to define and study creativity with the systematic tools of modern psychology was conducted by Lewis Terman at Stanford University. Terman began with the assumption that "genius" and "high intelligence" were the same thing. Starting in 1921, he launched a landmark "study of genius," which was continued after his death in 1956.

Terman's interest in giftedness began very early in his life. He was born in Indiana in 1877, the eleventh of fourteen children. His father was a farmer who loved books, but who was not otherwise distinguished. Little Lewis was small, red-haired, and not particularly athletic, so he was often teased by other boys. However, he was bright. Six months after he started school at age five, he was so precocious that he skipped to the third grade. His is a true Horatio Alger story. He had an intense desire to learn and to understand how the human mind works.

Lewis Terman

He also wanted to obtain a high level of education, at a time when most farm boys only completed eighth grade.

Impecunious and struggling with poor health caused by tuberculosis, he nonetheless rose to the top. He entered Indiana Central Normal College at fifteen, was teaching in a one-room country school at seventeen, and had obtained three baccalaureate degrees and logged many years of teaching by age twenty. In 1905 he obtained a Ph.D. from Clark University in Worcester, Massachusetts—a mecca at the time for studying the developing discipline of psychology. To finance his education he reluctantly borrowed $1,200 from his father and brother, a huge sum in that era. He promptly repaid them once employed as a teacher or professor, even though struggling to support a wife and family.

Having demonstrated early precocity himself, he was eager to test the then widely held belief that children who seemed to be very bright when young were likely to experience an intellectual and social decline when they became adults. The precocious child was considered to be abnormal, and those who believed that slower maturation led to a better long-term outcome in life followed the slogan, "Early ripe, early rotten." In short, genius in children was not a good thing. Terman's first study, completed as part of his doctoral degree from Clark, was conducted before IQ tests even existed. That study, which simply compared two contrasting groups of bright and dull children, did not confirm the "Early ripe, early rotten" hypothesis. He began to plan a long-term, longitudinal study of the life history of children who were considered to be highly gifted "geniuses."

Interest in measuring mental traits had soared in the pre–World War I era. Working in Paris in the early twentieth century, Alfred Binet had created the first tests of intelligence, inspired partly by his observation of how the minds of his two daughters bloomed and developed. Although trained as a lawyer, he was fascinated with how the human mind worked and devoted his life to studying human cognition. Without any formal training, he became one of the pioneering figures in the relatively new and growing field of psychology. His interest in finding a way to measure intelligence was further piqued when he served on a commission to evaluate the education of children considered to be retarded. The commission hoped to improve these children's education by identifying those who were capable of learning if given adequate stimulation. To do so, Binet created a group of simple tests of memory and reasoning. By 1910 he had begun to use these tests to make an important breakthrough: to take into account that a child's knowledge steadily changes and increases with age. He tested large numbers of children and determined how the average child performed on his tests at progressively older ages. That average for each age was the "mental age." He could then create a measure of intelligence, the "intelligence quotient" or IQ, by relating mental age, as measured by his tests, to actual chronological age by using a ratio. Intelligence Quotient equals mental age over chronological age, or IQ = MA / CA, with the result multiplied by 100. If you are ten years old, and your mental age is measured as equivalent to that of the average fifteen-year-old, your IQ is 150. This basic concept, formulated by Binet a century ago, is still the basis for measuring traditional IQ scores.

Meanwhile Terman had headed to California after completing his Ph.D., because his doctors had advised him to live in a milder climate than that of the Midwest. After teaching at the Los Angeles Normal School (which would eventually become UCLA), he joined the faculty at Stanford in 1910, where he ultimately became chairman of the Department of Psychology in 1922. Terman was a natural to follow up on the ideas of Binet, given his own interest in creativity, genius, intelligence, and giftedness. Working with several of his students at Stanford, Terman developed an English adaptation of Binet's test. He tested approximately 1,000 students with this new English version and published a summary of his research as a book, *The Measurement of Intelligence,* in 1916. A modest man, Terman did not name the test after himself. Instead, honoring the contributions of his Stanford students as well as Binet, he called it the Stanford Binet test. In that somewhat more altruistic and less greedy era, he also provided that the royalties earned by this test were to be used for psychology research, not for his own personal gain.

Terman's work at Stanford was interrupted by external events, however. The United States was becoming involved in World War I, and the military had a significant interest in measures of intellectual ability for several reasons. First, many American volunteers and inductees were found to be either illiterate or poorly educated, and thus ill equipped for a military role. Second, warfare was becoming more technologically complex, and it was important to have methods of assessing volunteers and inductees that would help predict their future success: who should become an officer, who should be assigned to handling more complex technical responsibilities, or who should be given simpler and less demanding support work. Terman had to put his long-term study of genius on hold. The United States entered World War I on April 6, 1917, and Uncle Sam needed Terman's talents as a psychologist and psychometrician. In the summer of 1917 a group of psychologists met and began the task of developing group tests that could be quickly administered to large numbers of people to obtain a good general assessment of their mental abilities. Terman became a prime mover in developing these standardized written tests to measure mental aptitude. Under his leadership, the Army Alpha and Beta tests were created. The Alpha was designed to test inductees who were literate. The Beta was for those who did not know how to read. The Alpha, given to more than 1.75 million men, was found to be enormously helpful in assigning individuals to the most appropriate rank and type of work within the military framework. Terman worked in Washington in the Surgeon General's office during the war years, first as a civilian and later as a major. The Army Alpha test was so robust that it was used again when the United States entered World War II.

After the war ended late in 1918, Terman was able to return to Stanford to pursue again his lifelong dream: to identify a group of children with high intellectual potential and to determine the long-term predictive value of a high IQ. He began his landmark study in 1921 by examining a cohort of children born around 1910 whose IQ scores ranged from 135 to 200, with an average of 151.5 for the boys and 150.4 for the girls. These subjects, who came to be known as "the Termites," were followed for more than 70 years. Terman continued to supervise the study until his death in 1956. It was subsequently carried on by many of his protégés, including his own son. The published results ran to six volumes.

The originality of Terman's study rests in its conceptualization. Not content simply to measure changes in intellectual scores, he instead devised a variety of other measures to examine nearly every aspect of the lives of these gifted children. He and his co-workers recorded information about the intellectual successes of relatives, body size measurements, medical history and examinations, educational history and achievement, personality testing, and intellectual and recreational interests. After the initial assessment, the evaluation continued systematically throughout each subject's lifetime and included physical and emotional development, occupational achievements, marital adjustment, and emotional and physical health. Very early, the study overturned the "Early ripe, early rotten" hypothesis, confirming Terman's doctoral research. It disproved the stereotype of the precocious child as scrawny, emotionally fragile, and socially inept. Terman's "geniuses" were typically above population norms or control group values. They were stronger, physically healthier, and more successful both economically and socially. Although psychiatric illnesses were not evaluated in a sophisticated way, the "Termites" appeared to have a good level of psychological health and reasonably good family adjustment.

Although it is intrinsically interesting because of its imaginative and pioneering design, the Terman study of genius is also intriguing because it sheds light on the relationship between intelligence and creativity. Some of Terman's "geniuses" did make notable and creative contributions. Robert Oppenheimer, for example, is known to have been a "Terman genius." In general, however, as the cohort matured, its members did not produce a significant number of creative individuals. There were very few successful writers, musicians, artists, innovative scientists, or creative mathematicians among them. The profile of the "Termites" at midlife (obtained between 1950 and 1952, when they were in their early forties) indicates a reasonable level of professional and material success, but not a striking level of creativity. Despite their high IQs, there were no Nobel Prize winners in the group. In fact, William Shockley and Luis Alvarez, both born in California during the time frame within which Terman's "geniuses" were recruited, did not make the cut—but they did become Nobel laureates in physics.

A summary of the occupations of the "Termites" at midlife shows that 45.6 percent were professionals, 40.7 percent managerial, 10.9 percent in retail business, 1.6 percent in agriculture, and 1.2 percent in semiskilled occupations. Of the entire 757 available for follow-up during midlife, only a few were conspicuously creative: two successful writers and one Oscar-winning film director. A few others in the group also displayed creative interests and some indirect successes, such as writing or painting as an avocation. Perhaps because Terman's original

IQ testing placed a high value on verbal skills, the majority who had creative interests or hobbies had chosen literary pursuits.

Conceived in the era during which intelligence testing was born, the Terman study of genius provided the first clear scientific evidence that genius (in the sense of creativity) was not the same as a high level of intelligence. Although the six-volume series produced by the study is collectively referred to as *A Genetic Study of Genius,* the individual volume titles indicate that Terman and his colleagues recognized this distinction between intelligence and creativity fairly early on. Volume 2 is entitled *The Early Mental Traits of 300 Geniuses,* but subsequent volumes have titles such as *The Gifted Child Grows Up: Twenty-five Years' Follow-up of a Superior Group* (volume 4) and *The Gifted Group at Mid-Life: Thirty-five Years' Follow-up of the Superior Child* (volume 5).

The distinction between intelligence and creativity has been supported by many later studies. For example, Roger MacKinnon, another California psychologist who studied creativity in the 1950s and 1960s, examined architects using a variety of measures, including intelligence. He divided the architects into three groups on the basis of creativity (highly creative, somewhat creative, and not creative). All three groups earned nearly identical scores on several different IQ tests, with an average of around 120 for each group. Thus those who were highly creative were not more intelligent than the less creative. Intelligence is somewhat related to creativity, but it is also different.

Much more has been written on this topic over the years. Although a universal consensus has never been achieved, there is probably a general agreement that a certain level of intelligence is needed in order to make a creative contribution of some kind. But at some point, another kind of mental/brain faculty kicks in, and it is this faculty—creativity—that permits some people to write novels, equations, sonnets, or symphonies.

Creativity and Society: Who Decides?

An old philosophical chestnut, fun to debate over dinner or over a drink, is the question: If a tree falls in a forest and no creature is around to hear it, does it actually make a sound? (A somewhat waggish and misogynistic contemporary version of this question runs: If a man speaks in the desert where no woman can hear, is he still wrong?)

In the context of the definition of creativity, we have a similar question. Aspiring writers, musicians, or artists work tirelessly to produce works they believe to express high levels of creativity, and yet frequently their work goes unpublished, unpresented, or unhung. Must we require that a creative person, and creative work, win external confirmation by publishers, critics, and other arbiters in order to judge the presence of genuine creativity?

Many real-life examples lead us to ponder this question. For example, Emily Dickinson, the "Belle of Amherst," wrote hundreds of poems during her lifetime that are striking in their originality of thought and their intensity of feeling. Most were not even published until after her death, and her works only very slowly gained the widespread critical acclaim and appreciation that they enjoy today. When did the act of creation occur? When she was actually writing the poems? Or only after they were discovered, published, and admired by society? Vincent van Gogh produced hundreds of paintings throughout his life. Yet no one, except a few friends, purchased any of his paintings, and he died an apparent failure. Only later did critical acclaim make his work widely sought after, and now his paintings sell for millions of dollars when auctioned at Sotheby's or Christie's. Most of John Donne's songs and sonnets, satires, and religious and secular love poems circulated in a handwritten underground form during much of his life. For three centuries they remained largely underground and appeared infrequently in anthologies until the early twentieth century, when T. S. Eliot rediscovered the metaphysical poets and held them up as ideal models of what poetry should be like.

Most of us would argue, I think, that these individuals, and many others who have gone unrecognized during their lifetimes or whose popularity waxes and wanes, should be defined as exhibiting creativity at the time that they were creating.

Others, however, have maintained the importance of using external standards to define the existence of "true creativity." Perhaps the strongest argument has been put forth recently by the psychologist Mihaly Csikszentmihalyi. He argues that:

> Creativity, the kind that changes some aspect of the culture, is never only in the mind of a person. That would, by definition, *not* be the case of cultural creativity. To have any effect, the idea must be couched in terms that are understandable to others, it must pass the muster with the experts in the field, and finally it must be included in the cultural domain to which it belongs.

He expands this position by suggesting that creativity can only be defined on the basis of the interrelationship of three components: the domain, the field, and the person.

A *domain* is an area of knowledge, such as mathematics, that forms a component of what we refer to as culture. The *field* consists of people who are gatekeepers to the domain. The field is comprised of critics, collectors, museum curators, journal reviewers, or funding agencies. As for the role of the *person,* Csikszentmihalyi states:

> Creativity occurs when a person, using the systems of a given domain such as music, engineering, business, or mathematics, has a new idea or sees a new pattern, and when this novelty is selected by the appropriate field for inclusion into the relevant domain.

A key component of this concept is that all three aspects are required for true creativity to become manifest: that creativity must create something new (typically transforming or enlarging a domain or creating a new one), and that the final arbiters as to what is creative must come from the field responsible for the domain. This somewhat novel model would therefore suggest that Dickinson, van Gogh, and Donne did not really become creative until the field decided that they were. Csikszentmihalyi refers to this as a "systems model." Using this model, he suggests that Gregor Mendel, who worked out the basics of modern genetics by cross-breeding peas, was not creative during his own lifetime. Mendel and others like him such as Copernicus, who were linked closely to the Church, kept their work secret and unpublished for fear of religious opprobrium. Mendel, Csikszentmihalyi argues, did not become creative until fifty years after his death, when people discovered his findings and began to put them to use in experimental genetic studies.

Csikszentmihalyi's distinctions and definitions have made a significant contribution to research on creativity. They have added rigor to psychological criteria for defining creativity by emphasizing the importance of objective evaluation. This is a valuable concept in a field that is still grappling with definitions and boundaries.

What Is Creativity?

We can define and conceptualize creativity in many ways. Boundary issues, such as the distinction between intelligence and creativity, must be considered. Both in popular language and in the historical literature, the terms "genius," "gifted," "talented," and "creative" are often used interchangeably. There are boundary issues with domains as well. Both in popular culture and in research studies, one sometimes senses a presumption that creativity occurs primarily in the arts and humanities—literature, music, dance, or visual arts—with little recognition that creativity is crucial for other fields as well, such as biology, mathematics, physics, chemistry, earth science, and engineering. And then there are judgment issues. Who decides that a creative product is genuinely creative, as opposed to not really very original, or as simply odd or idiosyncratic?

A resurgence of interest in creativity has occurred among psychologists over the past several decades. Consequently these issues have been discussed in detail in a variety of excellent books and articles. The authors are not in perfect agreement, but their work reveals enough common themes that I may formulate a definition of creativity for use in this book. This is essentially the definition that I used when I began my own research on creativity in the 1970s, and it still seems reasonable today.

One essential component of creativity is *originality*. Creativity involves perceiving new relationships, ways of observing, ways of portraying. These novel relationships might be found in nature and expressed in new natural laws, or expressed in a product such as a novel or a poem. Creativity is not limited to particular domains, such as the arts. Creativity in the sciences is also a fascinating topic.

A second component of creativity is *utility*, very broadly defined. For example, it is possible to conceive of something novel—such as a car without wheels—that has no creative value at all. The concept of utility must be broadly defined, however, because creativity in the arts is not always obviously useful. Its utility resides primarily in its ability to evoke resonant emotions in others, to inspire, or to create a sense of awe at what the human mind/brain can achieve.

A final component of creativity is that it must lead to a *product* of some kind. That is, creativity requires the *creation* of something. It may be helpful to think of creativity as comprised of three components. It begins with a *person*. That person then addresses a problem or seeks out a good question or conceives a new way of perceiving or conceptualizing, using a creative cognitive *process*. How that process occurs is a fascinating topic in cognitive neuroscience. When the process is complete—the problem is solved, the question answered, the work finished—there is a *product*. Person, process, product . . . These components may occur linearly, iteratively, or simply mysteriously.

Asking about how the individual brain moves through a creative process to produce a sonnet or a song or an equation is one of the most fascinating questions that we can contemplate. To answer it is to explain how we human beings, beginning with those early hunched and hairy prehistoric ancestors, have managed to wrestle ourselves out of dark caves and into a world ablaze with the brilliant fire and light of creative genius.

2

IN SEARCH OF XANADU

Understanding the Creative Person and the Creative Process

It was a miracle of rare device,
A sunny pleasure dome with caves of ice!
  A damsel with a dulcimer
  In a vision once I saw:
  It was an Abyssinian maid,
  And on her dulcimer she played,
  Singing of Mount Abora.
  Could I revive within me
  Her symphony and song,
  To such a deep delight 'twould win me,
That with music loud and long,
I would build that dome in air,
That sunny dome! Those caves of ice!
And all who heard should see them there,
And all should cry, Beware! Beware!
His flashing eyes, his floating hair!
Weave a circle round him thrice,

And close your eyes with holy dread,
For he on honey-dew hath fed,
And drunk the milk of Paradise.

—Samuel Taylor Coleridge, *Kubla Khan: Or, A Vision in a Dream*

How do human beings actually create the novel, beautiful, and useful ideas, images, and other products that are the hallmark of creativity?

Deceptively simple as thus stated, this is in fact a very complex question.

Philosophers, psychologists, and psychiatrists have been preoccupied with exploring answers to this question for many years. Ultimately, neuroscience will provide new answers, showing how the creative product arises from a process that occurs in an individual human brain. That process is neither easy nor obvious. Like the "First Creation" described in the book of Genesis, it evokes an aura of wonder, mystery, power, and even divinity. Although we shall not achieve a definitive answer to this complex question, we can drill down deeply and find answers that come increasingly close to understanding creativity at its multiple levels—the person, the process, and the neural events that lead to it.

Coleridge's Dream of Xanadu

One of the most famous descriptions of the creative process was written by the nineteenth-century Romantic poet Samuel Taylor Coleridge. He has described how he came to write what many people think of as his finest poem, *Kubla Khan: Or, A Vision in a Dream,* when he was just beginning his poetic career at age twenty-five. The poem, which remains a fragment, describes how Emperor Kubla Kahn created a magnificent palace in a place that Coleridge called Xanadu, encircled by a wall. Although the poem is ostensibly about building a palace, the creation of the palace is clearly a metaphor that conveys the turbulent processes that often underlie creative activity. It begins with lines that create a sense of power, grace, and delight, but also holiness, infinity, and darkness:

Samuel Taylor Coleridge

In Xanadu did Kubla Khan
A stately pleasure-dome decree:
Where Alph, the sacred river, ran
Through caverns measureless to man
   Down to a sunless sea.

The middle lines from the poem describe the actual act of creation as a violent and energetic natural process that seems almost beyond human control. They also convey the risks and threats that the creative person must face—the ancestral voices prophesying war.

And from this chasm, with ceaseless turmoil seething,
As if this earth in fast thick pants were breathing,
A mighty fountain momently was forced:
Amid whose swift half-intermitted burst
Huge fragments vaulted like rebounding hail,
Or chaffy grain beneath the thresher's flail:
And 'mid these dancing rocks at once and ever
It flung up momently the sacred river.
Five miles meandering with a mazy motion
Through wood and dale the sacred river ran,

Then reached the caverns measureless to man,
And sank in tumult to a lifeless ocean:
And 'mid this tumult Kubla heard from far
Ancestral voices prophesying war!

The poem concludes with the lines that begin this chapter. They suggest that creation—be it of a stately pleasure dome or of a poem or of a mathematical equation—is a complex and awe-inspiring process that contains elements of delight, but also fear, holiness, and a mystical inspiration. The road to Xanadu is a paradoxical path. On the one hand, it leads to a sunny pleasure-dome. On the other hand, the pleasure dome contains caves of ice!

Coleridge has described to us how he came to write this metaphorical description of the creative process. His description is enlightening, and it sets up themes that will be echoed repeatedly throughout this chapter. Troubled by ill health, Coleridge had retreated to a lonely farmhouse on the southern moors of England. He had been prescribed an anodyne (presumably opium) to relieve his symptoms. Under its influence, he fell asleep reading a description of Kubla Khan's palace in a book called *Pilgrimage,* by Samuel Purchas. Coleridge describes what subsequently happened:

> The Author continued for about three hours in a profound sleep, at least of the external senses, during which time he has the most vivid confidence, that he could not have composed less than from 200–300 lines; if that indeed can be called composition in which all the images rose up before him as *things,* with a parallel production of the correspondent expressions, without any sensation or consciousness of effort. On awakening, he appeared to himself to have a distinct recollection of the whole, and taking his pen, ink, and paper, instantly and eagerly wrote down the lines that are here preserved. At this moment he was unfortunately called out by a person on business from Porlock, and detained by him above an hour, and on his return to his room, found, to his no small surprise and mortification, that though he still retained some vague and dim recollection of the general purport of the vision, yet, with the exception of some eight or ten scattered lines and images, all the rest had passed away like the images on the surface of a stream to which a stone had been cast, but, alas! without the after restoration of the latter.

Coleridge managed to scrape together a few more lines, primarily lamenting the loss of the phantom world created during his altered state of reverie. He chose to treat as a complete work only the fifty-four lines that he wrote semi-automatically after awakening from his sleep. Although we might mourn the loss of the remaining 150 to 250 lines that he wrote during his "vision in a dream," for many of us *Kubla Khan* is a nearly perfect work of art as we currently know it.

Coleridge's description of his road to Xanadu, both in the poem itself and in his account of how he came to write it, raises many questions about the nature of creativity.

First, how important are personal factors—things such as experiences or lifestyle that are specifically intrinsic to a given individual at a given time? At the time he wrote *Kubla Khan,* Coleridge was not a particularly likely person to become a famous poet or to write one of the greatest poems in the English language. He was a college dropout. Although admired among his circle of friends because of his verbal agility, he was also a mystical dreamer, who had been making plans to leave Britain and found a socialist "Pantocracy" in the United States. He had significant health problems, which ultimately led to chronic use and abuse of opioids throughout his life. But these influences in fact converged on the southern moors of Britain in 1797 and led Coleridge to produce what is perhaps his greatest poem. His subject matter was fortuitous: he happened to be reading a book about the great Emperor. Very importantly, at the time he was reading, his brain was experiencing the dreamy sedation caused by opioids. Released from rational control and censorship, he spontaneously and seemingly instantaneously produced a convergence of images and metered words that constituted an entire poem.

Second, how much is creativity under conscious control? To what extent do unconscious processes predispose to the creation of a poem or an idea? Alternatively, how important is careful preparation, logical planning, and detailed thinking-through of a sequence or a topic in advance of the act of creation? By all accounts, *Kubla Khan* was literally created as "a vision in a dream," which was later recalled verbatim. It was not a consequence of any conscious effort. In fact, when Coleridge attempted to finish the poem using conscious effort, he failed completely. We have to ask how typical this is, and what other writers, artists, mathematicians, musicians, or scientists have to say about how they get their best ideas. How important is reason? How important is inspiration?

Third, how much is creativity influenced by a milieu or a social environment? Coleridge wrote this poem at a crucial time in his life. Although he had done well in school during earlier years, he did not complete his studies at Cambridge University. Preoccupied with a visionary idealism, he was living a drifting and somewhat aimless life. Fortunately, when he was about twenty-five, he met the other great Romantic poet of his generation, William Wordsworth, who befriended him and encouraged him to have a career as a poet. This career was launched in 1798, when Wordsworth and Coleridge jointly published *Lyrical Ballads,* containing such timeless poems as Coleridge's *The Rime of the Ancient Mariner* (and *Kubla Khan*), and Wordsworth's *Tintern Abbey.* These two innovative poets were also working at a pivotal time in the history of literature and the history of ideas. After the dominance of the rigid and rule-bound Neoclassical era of the eighteenth century, the world was ready for a change. The era of Romanticism was about to be launched, and Wordsworth and Coleridge were among the founders of this movement. *Kubla Khan, The Ancient Mariner,* and *Tintern Abbey* would be wonderful poems in any era, and they are still much admired in our spare postmodern twenty-first century. But no one could or would write them now, nor would *The New Yorker* or Knopf show any willingness to publish them. (A sad comment on twenty-first century publishing, but that's a topic outside the scope of this book.)

The Scientific Study of Creativity

Efforts to conduct scientific studies of creativity in a systematic manner began in the mid-nineteenth century, with the work of individuals such as the Italian psychiatrist Cesare Lombroso and the British scientist and naturalist Francis Galton. Galton, in particular, intro-

J.P. Guilford

duced new ways to measure the degree to which people vary in their levels of "genius." In fact, he invented basic concepts that helped create the field of statistics, which now form the basis for a great deal of psychological and psychiatric research. Lewis Terman, our friend from the previous chapter, was a careful student of Galton and thought of himself as carrying forward Galton's work. (Galton and Lombroso will receive more attention in chapter 4.)

Despite these major contributions, however, the study of creativity could hardly be described as a flourishing scientific field during the nineteenth and early twentieth centuries.

This vacuum was famously noted by yet another prominent psychologist from California in the 1950s and 1960s, J. P. Guilford. As president of the American Psychological Association in 1950, Guilford gave a presidential address that was a landmark in the history of the study of creativity. He made a clarion call to psychologists to undertake studies of what he regarded as one of humanity's most important traits: the ability to generate novel ideas. In his address, he noted that fewer than 0.2 percent of the entries in *Psychological Abstracts* focused on creativity. He argued that this important mental trait, which had contributed so much to the evolution of human society, should become a major field within the domain of psychology.

His plea did not pass unnoticed. During subsequent decades, many psychologists began to conduct well-designed research studies in an attempt to understand the nature of creativity. A variety of methods and tools were developed. As this work evolved, psychologists began to bump into several critical issues.

One of these was how creativity should be conceptualized. Is it a special gift given to only a few geniuses, or is it something that all human beings share? Stated in the language of statistics (a crucial tool for psychologists and psychometricians), is it a "continuous dimension" or a "discrete category"? This is an important distinction.

Continuous or dimensional traits are phenomena that people have in common, but for which there are individual differences—things such as height, weight, intelligence, or visual acuity. Many of these traits have (in statistical language) a "normal distribution," with most of the population scattered around the middle—or representing "the average"—while a few people are at either end of the curve and represent extreme examples (the very tall or small, the very smart or unintelligent, and so on).

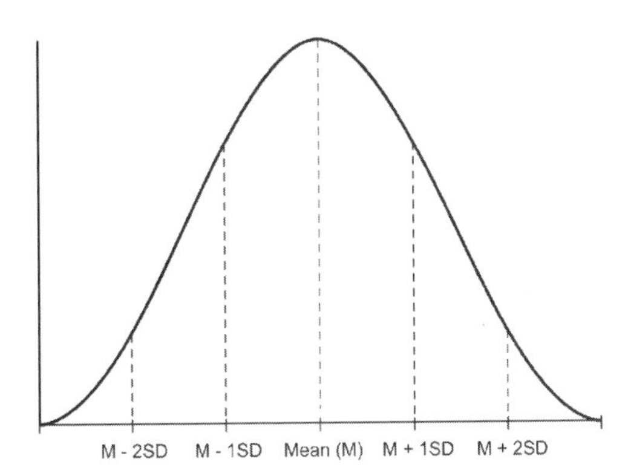

Figure 2-1. An Ideal Normal (Gaussian) Distribution
1 = one standard deviation (SD), 2 = two standard deviations. 95% of the population is within two standard deviations from the mean on either side of the mean. Approximately 2/3 are within one standard deviation. IQ has a normal distribution in the population.

If creativity is a mental capacity that is distributed on a continuum, then it could be studied in much the same way that Terman studied intelligence. Psychologists would just need to figure out how to do for creativity what Terman did for IQ—design good tests to measure it. Then, as Terman did, people could be identified who are at the far end of the distribution (usually the top 1 percent) and who could therefore be defined as highly creative.

If this approach is successful (and if its underlying assumptions about creativity as a normally distributed trait are true), then the implications for education are significant. Creatively gifted children could be identified through standardized psychometric tests designed to evaluate creativity and to identify those children who are at the high end of the normal distribution. Once identified, their creativity could be nourished and enhanced, thereby providing human society with a greater resource of talent in many fields of science and the humanities. (It is probably no accident that many of the current creativity researchers come from a background in education.) Identifying more creative individuals might also be useful for other purposes, such as identifying candidates to serve in Special Forces in the military or screening candidates for promotion to higher executive levels in business.

An alternative point of view, however, is that creativity is not a dimensional trait. Rather, it is a characteristic of rare, unique, and unusual individuals. Instead of being at the extreme end of a continuous "bell curve" that represents the normal population (as high intelligence is), the capacity for creativity is a discontinuous trait or group of traits that occur uniquely in a few extremely gifted individuals. Figure 2-2 illustrates the theoretical distribution of this type of nondimensional trait. Quite literally, creativity is a "gift," divinely bestowed from God or occurring as a near-miraculous biological or social accident (depending on a person's worldview). According to this point of view, creativity can be studied successfully only by focusing on highly gifted people—people who have an established track record of being creative in some field, such as music, art, literature, architecture, mathematics, or science.

I believe both views are right. I think of the first kind as "ordinary

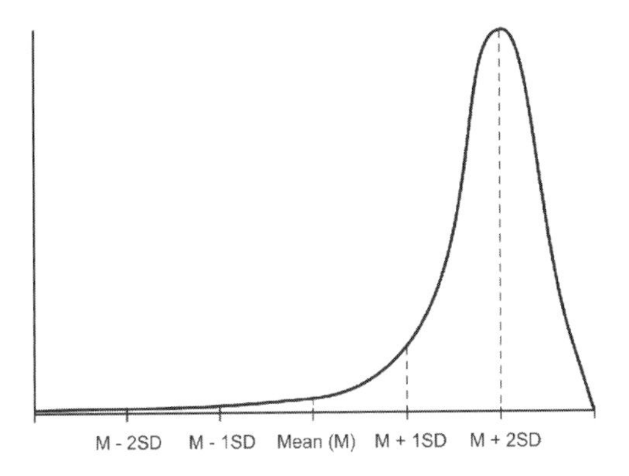

Figure 2-2. A Theoretical Depiction of Skewed Distribution
Highly gifted individuals may fall in an area 2 standard deviations above the mean and represent a very distinct and unusual group of people.

creativity," while the second is "extraordinary creativity." Ordinary creativity can be nurtured and enhanced in a variety of ways. Extraordinary creativity, or creativity in its purest form, is my main emphasis in this book, because examining the minds and brains of great geniuses is a fascinating endeavor that teaches us a lot about the creative process.

Psychologists who heard Guilford's call to study creativity also bumped into another critical issue. What is the best way to study creativity? To some extent, this second issue is related to the first: how creativity is conceptualized. One common method has been to emphasize the study of ordinary creativity. If one believes that creativity is a continuous trait, then the best strategy is to develop experimental tests of creativity and to apply them to large numbers of people, selecting out those who score high as representing "the creative person." Much as Terman did with his "geniuses," these people can then be examined in a variety of ways, such as reevaluating them at regular intervals to determine their long-term history of life success. One can also measure their IQs to examine the relationship between creativity and intelligence, or study them with other types of cognitive tests or physiologic measures.

Alternatively, one can concentrate one's efforts on the study of extraordinary creativity. If we believe that creativity is a rare and exceptional trait, then we should focus primarily on the study of people selected because of their history of creative success. This approach emphasizes the use of personal interviews and introspective accounts. Potentially, people who have extraordinary creativity could also be studied using other measures, including modern neuroimaging tools.

A final issue is what the student of creativity should actually study—that is, whether to examine personality traits, thought processes, or other kinds of measures. To some extent, the history of studies of creativity reflects the history of psychology in the twentieth century. Psychological research on creativity has used the methods that have been "cutting edge" in the field at a given time. The last half of the twentieth century was characterized by the development of tools to measure personality and cognition. Tests such as the Minnesota Multiphasic Personality Inventory (MMPI) and the Eysynck Personality Inventory were developed as psychologists began to study individual differences in personality. In parallel, the discipline of experimental cognitive psychology emerged, with the goal of determining how cognitive processes such as memory or attention can be measured and divided into more specific components such as memory encoding versus retrieval, or focused versus divided attention. A new subdiscipline, historiometrics, has been developed primarily by University of California psychology professor Dean Keith Simonton. This approach is a highly developed way of studying topics such as scientific productivity by using such measures as the number of times that a publication is cited. Many of these tools have been applied to the study of creativity.

The Creative Person

Guilford's call occurred at a fortuitous time. Just as national military interests had a significant influence on the study of intelligence after World War I, World War II had a similar effect on the study of creativity. The United States had just emerged from the terrible toll of World War II. The war had literally come to an end with a bang. The creation of the atomic bomb forced the Japanese to a sudden surrender, saving many lives of Americans, Australians, British, and even the Japanese themselves. Although in hindsight we now mournfully recognize how much the creative products of the human brain may proliferate and become ever more powerful in their potential to destroy, nevertheless the discovery and recognition of the power of atomic energy was also one of the great achievements of the twentieth century. It was also an example of how creative capacities could sometimes be enhanced when people worked together in a dedicated group. The Manhattan Project demonstrated that creative collaborations could be of value not only in pure science, but also in solving practical problems such as improving national defense. Therefore, the government developed an interest in investing in creativity research, and several psychologists were well funded to conduct such work.

Interviews and the Case-Study Method

The Case-Study Method emerged as an important research strategy during this time period. This method involves the study of people who have extraordinary creativity. People with extraordinary creativity can be identified in several different ways, in order to make sure that the selection is reasonably objective. Some investigators use the "jury approach": eminent leaders in a particular field are asked to provide a list of the most creative people in that field, and subjects are then recruited from the list provided by the "jury." An alternative approach is to select people who have received honors given to the most eminent individuals in a particular field, such as winners of Nobel Prizes or other prominent awards.

The value of this method is that it bypasses having to find a good way to measure creativity, which is not an easy task. Investigators who have used the Case-Study Method have typically either personally visited, interviewed, and tested their subjects, or they have brought them to their respective research institutions for a period of intensive evaluation.

Some excellent examples of the Case-Study Method include the work of Anne Roe, Frank Barron, D. W. McKinnon, John Drevdahl, and Raymond Cattell. Their work has yielded many interesting findings. For example, many of these psychologists took a closer look at the relationship between intelligence and creativity. Their results firmly nailed down the ideas suggested by Terman's study and suggest that we should think about this relationship in terms of what has been called the Threshold Theory. The Threshold Theory holds that above a certain threshold, intelligence and creativity are not closely related to one another. (In the language of psychometrics, they are not correlated.)

Subjects in these studies were subjected to a variety of IQ tests and on average were found to have IQs in the 120–130 range. That is, they were definitely highly intelligent, but they did not achieve extreme scores that would be consonant with their extremely high levels of creativity. The general conclusion is that most creative people are smart, but they don't have to be extremely smart. An IQ around 120 is generally good enough. The one exception, interestingly, is Anne Roe's study of scientists, which included biological scientists, physicists, and social scientists. Her tests of intelligence were divided into verbal, spatial, and mathematical domains. While the social scientists and the biologists were above average on the mathematical tests, the physicists simply refused to take them because they were too easy!

The Creative Personality

The case study work of Roe and others included both personality tests and extensive interviews. It has revealed a personality profile of what the creative person is like. Although some personality differences may be related to a particular form of creativity (e.g., science vs. literature), a surprisingly large number of personality traits are shared across fields. The following description of the creative personality is drawn from the work of the pioneers in the Case-Study Method, as well as from my own work interviewing creative writers from the University of Iowa Writers' Workshop, which I'll describe in more detail in chapter 4.

Personality traits that define the creative individual include openness to experience, adventuresomeness, rebelliousness, individualism, sensitivity, playfulness, persistence, curiosity, and simplicity.

Creative people tend to approach the world in a fresh and original way that is not shaped by preconceptions. The obvious order and rules that are so evident to less creative people, and which give a comfortable structure to life, often are not perceived by the creative individual, who tends to see things in a different and novel way. This openness to new experience often permits creative people to observe things that others cannot, because they do not wear the blinders of conventionality when they look around them. Openness is accompanied by a tolerance for ambiguity. Creative people do not crave the absolutism of a black and white world; they are quite comfortable with shades of gray. In fact, they enjoy living in a world that is filled with unanswered questions and blurry boundaries.

Creative people enjoy adventure. They like to explore. As they explore, they may push the limits of social conventions. They dislike externally imposed rules, seemingly driven by their own set of rules derived from within. This lack of commonality with the rest of the world may produce feelings of alienation or loneliness. In addition, the lack of evident and obvious standards of perception or information may produce a blurring of the boundaries of identity or self, sometimes referred to in psychodynamic terminology as ego boundaries. This may be one of the traits that explains the higher rate of mental illness in creative people than in the general population, as described in chapter 4.

Paradoxically, the creative person's indifference to convention is combined with sensitivity. This may take two forms: sensitivity to what others are experiencing, or sensitivity to what the individual himself or herself is experiencing. Creative people tend to have high levels of both forms. Inevitably, this combination of pushing the edge and experiencing strong feelings can lead to a sense of injury and pain. Living on "the edge of chaos" may also be psychically dangerous, because approaching too close may even lead to "falling off" occasionally—into mental disorganization or confusion. This may also partially explain why creative people have higher rates of mental illness.

Nevertheless, creative people have traits that make them durable and persistent. Adventurousness and rebelliousness are often coupled with playfulness. The capacity to approach the world in a playful and even childlike manner adds an intermittently joyous tone to the life of the creative person.

Creative people also have an ability to persist in spite of repeated rebuffs. Persistence is absolutely fundamental, since creative people typically experience repeated rejection because of their tendency to push the limits and to perceive things in a new way. Young poets, biologists, playwrights, physicists, and other creative people all must experience being turned down, whether it is having their writings rejected or their grant applications denied funding. Creative people must retain the capacity to keep going, even in the face of very little external validation of their worth.

Creative people also tend to be intensely curious. They like to understand how and why, to take things apart and put them back together again, to move into domains of the mind or spirit that conventional society perceives as hidden or forbidden. Their curiosity also has a driven and energetic quality. Once absorbed in an idea or topic, they pursue it with a dogged intensity. In fact, a theme that runs through all the work done using the Case-Study Method is that creative people work very long hours, far longer than the average person's eight-hour day. Many of the great creators, such as Michelangelo, are notorious for working almost night and day. Creative people are also often perfectionistic and even obsessional. They must work on a topic, project, or idea until they "get it right."

These traits tend to be combined with a basic simplicity, defined by a singleness of vision and dedication to their work. In fact, much of the time, their work is really all that creative people care about.

The Creative Process

The second step in our journey to Xanadu is to examine the nature of the creative process. What actually happens, at the cognitive level, when a person creates something that is novel, useful, and beautiful? Historically, scholars have taken two approaches to this question. One emphasizes the importance of using objective measures of creativity. Its technique has been to devise experimental tests that can measure creativity, just as IQ tests were designed to measure intelligence. An advantage of this approach is that it can be applied to large groups of people, just as the Army Alpha intelligence test was. The second approach uses more subjective tools—and some would say is less scientific. This approach relies on introspective accounts from people who have extraordinary creativity, observing the mental processes that they have gone through during the process of creating. The strength of this approach is that it may come closer to the essence of the creative process, because it examines it in extraordinarily creative people.

Experimental Tests

Experimental tests to measure creativity are similar to IQ tests in their basic assumptions. First, they assume that creativity is a continuous or dimensional trait that all people share, and that people who test at the high end can be designated as "highly creative." A second assumption is that experimental tests can actually be devised to measure creativity accurately. Creativity tests have been used for several different purposes. One goal has been to measure levels of creativity in representatives of the general population and to determine what the people who achieve high scores are like. Another goal has been to use these tests to screen selected groups, such as candidates for astronaut training, in order to select out the more creative people.

The challenge facing this psychometric strategy has been to find convincing ways to measure creativity. To some extent, creators of creativity tests have to be creative enough themselves to "get inside" the minds of creative people and to understand how they work. If they can understand the creative process, then they can figure out ways to assess whether people have a capacity for creativity.

People have debated about what creativity really is and whether it can be measured at all. Within the framework of psychology and psychometrics, measures that seem convincing and plausible are referred to as having "validity." Several kinds of validity have been defined. One is "face validity." This simply means that the contents of a test "make sense" (in this case, as a measure of creativity) to reasonably knowledgeable people. A second type of validity is known as "construct validity." This means that the contents of the test derive from a plausible theory about what the nature of creativity is. A third type of validity is "predictive validity," which means that the test can predict something that is relevant to the concept of creativity, such as having future success in some area of life that requires creativity.

The psychologists who have designed creativity tests have usually assumed that a key component of the creative process is the capacity to show a thinking style that they have called *divergent thinking*. This kind of thinking is the ability to produce a large number of interesting, appropriate, and fluent responses to some kind of a probe, such as a question or a task. Creativity test designers have usually made a distinction between divergent thinking (which emphasizes coming up with many possible answers and is considered to be a hallmark of creative thought processes) and *convergent thinking* (which stresses finding a single right answer and is a hallmark of more conventional thought processes). A typical example of a probe that can be used to measure the ability to think divergently, from one of the standard creativity tests, is "How many uses can you think of for a brick?" Because the concept of divergent thinking forms the bedrock for most of the existing tests of creativity, the face and construct validity of these tests depends on how plausible the concept is.

Not surprisingly, Guilford himself was one of the earliest inventors of creativity tests. His tests were based on his theories about the structure of the intellect. These theories presaged the current views of Howard Gardner, Robert Sternberg, and others, who concur that intelligence is not a unitary construct but rather is comprised of multiple abilities.

Guilford developed a test of creativity called the Structure of Intellect (SOI) Test, which included a battery of subtests. This test was designed for use in the general population and was appropriate for use with school children. Guilford's SOI Test examined fluency of thought in both the verbal and the visual domains. For example, a verbal subtest might ask the subject to think of as many consequences as possible if people no longer needed to sleep. The visual component of the test might ask the subject to look at a group of different figural shapes, such as triangles or circles, and to classify them in as many different ways as possible. Many variants of creativity tests were developed by others, such as the Torrance Test of Creative Thinking (TTCT). These tests are still widely used in many settings to evaluate the potential for creativity. For example, some schools may use them to identify "gifted children" who should be fast-tracked, or the military might use them to identify people who have the problem-solving abilities needed to perform well in Special Forces.

The use of these tests is not without controversy, however. They have been criticized for many reasons. Despite appearing to be objective, their scoring is necessarily somewhat subjective. The entire construct of divergent thinking has also been criticized based on the view that ideational fluency is not the same as originality. Perhaps the harshest criticism, however, is that these tests may lack predictive validity. Several different research teams have administered such tests to groups of school children who were later evaluated in terms of their creative success. In general, those who achieved higher test scores did not necessarily have greater success in later life than low scorers. Because one of the major motivations for developing these tests was to identify individuals with future creative potential, enthusiasm for using them has been diminished.

The Case-Study Method and Introspective Descriptions of the Creative Process

Although less objective and quantitative than experimental tests, interviews and case studies of creative people can tell us a great deal about how creative individuals actually produce works of art or science. Hearing a creative person describe how he or she actually thinks and works probably illuminates more aspects of the creative process than can be obtained from experimental tests.

For example, several years ago I happened to sit next to the distin-

Neil Simon

guished American playwright Neil Simon on an airplane. For me, Simon has a degree of creativity reminiscent of Shakespeare's. First, he is extremely prolific, having written more than forty plays so far, many of which have been staged to great public acclaim. Second, although his plays have enormous popular appeal, they also are rich in content and ideas. Many of the characters in his plays have become cultural icons, such as Felix and Oscar of *The Odd Couple.* He has explored many themes, such as coming of age *(Brighton Beach Memoirs, Biloxi Blues, Broadway Bound)*, the relationship between men and women *(Last of the Red Hot Lovers)*, or old age *(The Sunshine Boys)*. Third, his output is varied. Some plays, such as *The Odd Couple,* are quite comic, while others such as *Chapter Two* probe sadder aspects of life.

Because we shared an interest in the nature of creativity, Simon and I spent several hours in an intense discussion about the nature of the creative process. I had already interviewed a large number of writers for my Writers' Workshop study, and so Simon's description of his own thought processes struck many familiar notes. But he described these mental processes in a manner that was particularly precise and apt. Four quotations from this interview nicely illustrate key components of the creative process.

"I slip into a state that is apart from reality."

In order to create, many creative people slip into a state of intense concentration and focus. In psychiatric terms, this could be described as a "dissociative state." That is, the person in a sense mentally separates himself from his surroundings and metaphorically "goes to another place." In ordinary language, the person might be said to be "no longer in touch with reality." In a subjective sense, however, the creative individual is moving into another reality that is actually more real. Although the person outwardly appears conscious but "lost in thought," this reality is similar to an unconscious state, a place where words, thoughts, and ideas float freely, collide, and ultimately coalesce. This experience of "disengagement," "intense focus," or "being in another place" is probably somewhat analogous to the altered states described by the great mystics. Once in that "other reality," the creative person may remain for hours on end, living in a world that is a mixture of nebulous floating concepts and forms that are gradually turned into an object or idea that ultimately becomes the creative product, be it a play or a mathematical formula. This capacity to focus intensely, to dissociate, and to realize an apparently remote and transcendent "place" is one of the hallmarks of the creative process.

"I don't write consciously—it is as if the muse sits on my shoulder."

In general, creativity is not a rational, logical process. Although organization, structure, and planning are brought to bear in designing the overall shape of a play, a building, or a chemical synthesis, the *essence* of a creative product usually cannot be consciously planned or willed into existence. The notion of the muse, or the need for inspiration, is much more than a metaphor. Most creative people express in some way the idea that during the process of creation which occurs in "that place," they are simply expressing unconscious thoughts and processes. They are prone to make statements such as "I don't know where it comes from; it just happens." Neil Simon actually stated that he never knows how the particular play that he is writing is going to end until he finishes the last few scenes and bits of dialogue. He smiled and said, "If I knew how it was going to come out in advance, it would probably ruin the play, since I would be giving the audience clues that would tip them off and spoil the psychological suspense that builds to the final ending."

"My mind wanders—even when I talk."

Creative people are more prone to be flooded with ideas and thoughts, and also less likely to censor them, because these are the stuff of which their art or science is made. As we understand more and more about how the brain works in our current age of neuroscience, we are observing that highly creative people are likely to have brains that think somewhat differently. In the language of cognitive psychology, they are less likely to censor input either from within or from without. This is sometimes referred to as an altered "filtering mechanism." Within the creative individual, "the wandering mind" is experienced as producing a steady input of ideas that may be somewhat fragmented and formless. To the outside observer, the person may seem to move rapidly from one topic to another. For the creative person, this mechanism leads to heightened perceptions, greater sensitivity to external stimuli, and increased intensity of experience.

"I've always felt as if I'm invisible."

This may seem like a strange comment to come from a well-known playwright, but it actually makes sense. Creative people tend to be observers. Furthermore, they attempt to invalidate Heisenberg's Uncertainty Principle that "there is no observer outside the experiment." If invisible, they may observe more, and more accurately.

Creative people have the capacity to be disengaged and dispassionate observers. To others they may seem aloof, detached, or even coldhearted at times. To themselves, they often feel as if they are watching the rest of the world without others even knowing about it. This trait may seem to run counter to the flamboyance displayed by some creative individuals, who may appear to be seeking attention rather than invisibility. Nevertheless, many creative people, even when flamboyant, express a subjective sense that they can see into other people without being noticed. In a sense, they have an ability to spy on the universe.

Neil Simon's comments could be dismissed as irrelevant to our goal of understanding the nature of the creative process. After all, they represent the subjective perspective of a single person. Yet an accumulating body of similar observations from other highly creative people suggests that Simon's insights do indeed express some key features of the process.

Five Introspective Accounts

In letters, interviews, and books, many other highly creative individuals have described how their creative ideas have come to them. Because such people also represent extraordinary creativity in its purest form, reading their descriptions can give us unique insights into how the human mind and brain create thoughts and works that are beyond the reach of most ordinary human beings. The introspective accounts below represent samples from music, mathematics, science, and literature.

Wolfgang Amadeus Mozart

Let us begin first with a description by the prodigious musical genius Wolfgang Amadeus Mozart. The description is of a process not dissimilar to that described by Coleridge:

Wolfgang Amadeus Mozart

When I am, as it were, completely in myself, entirely alone, and of good cheer—say traveling in a carriage, or walking after a good meal, or during the night when I cannot sleep; it is on such occasions that my ideas flow best and most abundantly. *Whence* and *how* they come, I know not; nor can I force them. Those pleasures that please me I retain in memory, and am accustomed, as I have been told, to hum them to myself. If I continue in this way, it soon occurs to me how I may turn this or that morsel to account, so as to make a good dish of it, that is to say, agreeably to the rules of counterpoint, to the peculiarities of the various instruments, etc.

All this fires my soul, and, provided I am not disturbed, my subject enlarges itself, becomes methodized and defined, and in the whole, though it be long, stands almost complete and finished in my mind, so that I can survey it, like a fine picture or a beautiful statue, at a glance. Nor do I hear in my imagination the parts *successively*, but I hear them, as it were, all at once *(gleich alles zusammen)*. What a delight this is I cannot tell! All this inventing, this producing, takes place in a pleasing lively dream. Still the actual hearing of the *tout en-*

semble is after all the best. What has been thus produced I do not easily forget, and this is perhaps the best gift I have my Divine Maker to thank for. When I proceed to write down my ideas, I take out of the bag of my memory, if I may use that phrase, what has been previously collected into it in the way I have mentioned. For this reason, the committing to paper is quickly done, for everything is, as I have said before, already finished; and it rarely differs on paper from what it was in my imagination. At this occupation I can therefore suffer myself to be disturbed; for whatever may be going on around me, I write, and even talk, but only of fowls and geese, or of Gretel or Bärbel, or some such matters. But why my productions take from my hand that particular form and style that makes them *Mozartish,* and different from the works of other composers, is probably owing to the same cause which renders my nose so large or so aquiline, or, in short, makes it Mozart's, and different from those of other people. For I really do not study or aim at any originality.

Peter Ilych Tchaikovsky

Here is an introspective account from another musician, Peter Ilych Tchaikovsky, who composed two centuries later. Again, the similarities to the descriptions by Mozart and Coleridge are striking. Tchaikovsky also emphasizes the inspirational, supernatural, dreamlike, and unconscious nature of much of the creative process. He, like Coleridge, suggests that this process can be consuming and almost violent in its effects. And, like Coleridge but unlike Mozart, he laments how difficult it can be to reconstruct the composition if the thread of inspiration is broken off in the middle. Interestingly, he also emphasizes that although composition is not "a cold exercise of the intellect," the composer is not excused from sitting down and trying. If he works and waits long enough, the muse will eventually sit on his shoulder again.

Generally speaking, the germ of a future composition comes suddenly and unexpectedly. If the soul is ready—that is to say, if the disposition for work is there—it takes root with extraordinary force and rapidity, shoots up through the earth, puts forth branches, leaves, and, finally, blossoms. I cannot define the creative process in any other way than by this simile. The great difficulty is that the germ must appear at a favorable moment. The rest goes of itself. It would be vain to try to put into words that immeasurable sense of bliss which

Peter Ilych Tchaikovsky

comes over me directly a new idea awakens in me and begins to assume a different form. I forget everything and behave like a madman. Everything within me starts pulsing and quivering; hardly have I begun the sketch ere one thought follows another. In the midst of this magic process it frequently happens that some external interruption wakes me from my somnambulistic state: a ring at the bell, the entrance of my servant, the striking of the clock, reminding me that it is time to leave off. Dreadful, indeed, are such interruptions. Sometimes they break the thread of inspiration for a considerable time, so that I have to seek it again—often in vain. . . . if that condition of mind and soul, which we call inspiration, lasted long without intermission, no artist could survive it. The strings would break and the instrument be shattered into fragments. It is already a great thing if the main ideas and general outline of the work come without any racking of the brains, as the result of that supernatural inexplicable force we call inspiration. . . .

Do not believe those who try to persuade you that composition is only a cold exercise of the intellect. The only music capable of moving and touching us is that which flows from the depths of the composer's soul when he is stirred by inspiration. There is no doubt that even the greatest musical geniuses have sometimes worked without inspiration. The guest does not always respond to the first invitation. We must *always* work and a self-respecting artist must not fold his hands on the pretext that he is not in the mood. If we wait for the mood, without endeavoring to meet it halfway, we easily become indirect and apathetic.

Henri Poincaré

Poincaré's introspective account takes us to a different type of creativity, but again we see a surprisingly similar description of the nature of the creative process. Poincaré is known primarily as a mathematician, but his interests were very diverse. They also included physics and astronomy. He worked on many aspects of complex algebra and geometry, including differential equations, and invented the new mathematical field of topology. In this account he describes how he solved a series of mathematical problems over an extended period of time, interrupted intermittently by other unrelated tasks and by travel. Each time a component of the problem was solved, it came to him without conscious effort, although it was sometimes followed by conscious activity to work out the details. Thus, in this account of mathematical discovery, there is a resonance between an inspired nonrational state and a more rational state in which the details were elaborated.

For fifteen days I strove to prove that there could not be any functions like those we have since called Fuchsian functions. I was then very ignorant; every day I seated myself at my work table, stayed an hour or two, tried a great number of combinations and reached no results. One evening, contrary to my custom, I drank black coffee and could not sleep. Ideas rose in crowds; I felt them collide until pairs interlocked, so to speak, making a stable combination. By the next morning I had established the existence of a class of Fuchsian functions, those which come out from the hypergeometric series; I had only to write out the results, which took but a few hours.

Then I wanted to represent these functions by the quotient of two series; this idea was perfectly conscious and deliberate, the analogy with elliptic func-

Henri Poincaré

tions guided me. I asked myself what properties these series must have if they existed, and I succeeded without difficulty in forming the series I have called theta-Fuchsian.

Just at this time I left Caen, where I was then living, to go on a geologic excursion under the auspices of the School of Mines. The changes of travel made me forget my mathematical work. Having reached Coutances, we entered an omnibus to go someplace or other. At the moment when I put my foot on the step the idea came to me, without anything in my former thoughts seeming to have paved the way for it, but the transformations I had used to define the Fuchsian functions were identical with those of non-Euclidean geometry. I did not verify the idea; I should not have had time, as, on taking my seat in the omnibus, I went on with the conversation already commenced, but I felt a perfect certainty. On my return to Caen, for conscience' sake, I verified the result at my leisure.

Then I turned my attention to the study of some arithmetical questions apparently without success and without a suspicion of any connection with my preceding researches. Disgusted with my failure, I went to spend a few days at the seaside, and thought of something else. One morning, walking on the bluff, the idea came to me, with just the same characteristics of brevity, suddenness and immediate certainty, that the arithmetic transformations of indeterminate tenary quadratic forms were identical with those of non-Euclidian geometry.

Returned to Caen and meditated on this result and deduced the consequences. The example of quadratic forms showed me that they were Fuchsian groups other than those corresponding to the hypergeometric series; I saw that I could apply to them the theory of theta-Fuchsian series and that consequently there existed Fuchsian functions other than those from the hypergeometric series, the ones I then knew. Naturally I set myself to form all these functions. I made a systematic attack upon them and carried all the outworks, one after another. There was one however that still held out, whose fall would involve that of the whole place. All my efforts only served at first the better to show me the difficulty, which indeed was something. All this work was perfectly conscious. Thereupon I left from Mont-Valérien, where I was to go through my military service; so I was very differently occupied. One day, going along the street, the solution of the difficulty which had stopped me suddenly appeared to me. I did not try to go deep into it immediately, and only after my service did I again take up the question. I had all the elements and had only to arrange them and put them together. So I wrote out my final memoir at a single stroke and without difficulty.

Friedrich Kekulé

One of the most famous accounts of a dreamlike inspiration comes from the nineteenth-century German organic chemist Friedrich Kekulé. Kekulé had devoted much of his career to studying the nature of carbon bonding and the structure of benzene. As we have now seen so often about the process of creativity, Kekulé finally solved the problem while asleep and dreaming.

I was sitting writing on my textbook, but the work did not progress; my thoughts were elsewhere. I turned my chair to the fire and dozed. Again the atoms were gamboling before my eyes. This time the smaller groups kept modestly in the background. My mental eye, rendered more acute by the repeated visions of the kind, could now distinguish larger structures of manifold confor-

Friedrich Kekulé

mation; long rows sometimes more closely fitted together all twining and twisting in snake-like motion. But look! What was that? One of the snakes had seized hold of its own tail, and the form whirled mockingly before my eyes. As if by a flash of lightning I awoke; and this time also I spent the rest of the night in working out the consequences of the hypothesis.

Instead of thinking of chemical bonds as straight lines, Kekulé realized that bonds could create a shape that would be like a snake chasing its tail. The substance benzene was a circle, or what we now know as the benzene ring.

Stephen Spender

Our final example, comments by Stephen Spender, formerly British Poet Laureate, also emphasizes the importance of inspiration in the creation of a poem. By Spender's account, the inception of a poem begins with "a given line" ("une ligne donnée"), which comes as an effortless insight. Unlike Coleridge and other creative individuals we have considered so far, however, Spender also stresses that there must be an effortful process that amplifies on the "given line" and creates the remainder of the poem. He also emphasizes the importance of the poet's individualistic self-confidence, a quality of the creative personality described earlier in this chapter.

> Inspiration is the beginning of a poem, and it is also its final goal. It is the first idea which drops into the poet's mind and it is the final idea which he at last achieves in words. In between this start and this winning post, there is the hard race, the sweat and toil. Paul Valéry speaks of the "une ligne donnée" of a poem. One line is given to the poet by God or by nature, the rest he has to discover for himself.

Stephen Spender

> It is evident that a faith and their vocation, mystical intensity, sustains poets. There are many illustrations from the lives of poets to show this, and Shakespeare's sonnets are full of his expressions of his faith in the immortality of his lines. . . .
>
> Although it is true that poets are vain and ambitious, their vanity and ambition is of the purest kind attainable in this world, for the saint renounces ambition. They are ambitious to be accepted for what they ultimately are as revealed by their inmost experiences, their finest perceptions, their deepest feelings, their uttermost sense of truth, in their poetry. They cannot cheat about these things, because the quality of their own being is revealed not in the noble sentiments which their poetry expresses, but in sensibility, control of language, rhythm in music, things which cannot be attained by a vote of confidence from an electorate, or by the office of Poet Laureate.

I find introspective accounts such as these to be fascinating. They give us a direct insight into how the creative mind and brain frequently work during the process of achieving a creative product. Thought often moves swiftly and multidimensionally. The solution occurs in a flash. It may occur after a "rest period," during which ideas lie fallow, and then suddenly take root and sprout. It is experienced as "inspiration"—Neil Simon's muse sitting on his shoulder.

If this is the creative process, and if this process ultimately arises from the human brain, how does the brain create that process? Can the brain understand itself at all?

Follow me into the next chapter, where we'll continue on our search for Xanadu, to explore the neural mechanisms of creativity.

3

REACHING XANADU

How Does the Brain Create?

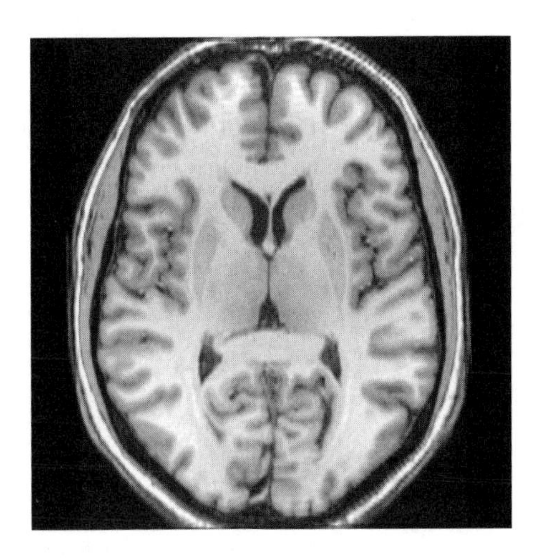

When to the sessions of sweet silent thought,
I summon up remembrance of things past . . .

—William Shakespeare, Sonnet 30

Just as Guilford complained fifty years ago about psychology's lack of attention to creativity and challenged his field to rectify the problem, one might make the same complaint today about the scarcity of neuroscientists studying creativity. The late twentieth century has been characterized by the development of our new discipline, neuroscience, which integrates a multiplicity of methods for studying the brain, cognition, and even personality. These methods include everything from molecular and cell biology to cognitive neuroscience. Until very recently, studies of the neural basis of creativity have been all too rare. This is a strange and vexing situation because, by examining the neural processes underlying thinking in general and creativity in particular, neuroscience can potentially take us to the very essence of what creativity is.

There is something magical about journeying to one of the last scientific frontiers—the human brain—in order to examine how its most precious creative achievements arise. Many neuroscientists, few of whom would characterize themselves as "romantic" in the Coleridgian sense, nonetheless understand the spirit and meaning of the final lines of *Kubla Khan:*

> . . . close your eyes with holy dread,
> For he on honey-dew hath fed,
> And drunk the milk of Paradise.

Most of us feel that we too are in search of Xanadu. We are privileged to be able to get up each day in order to study the most interesting organ in the most interesting creature on earth, and to ask the most interesting questions ever asked by science. Each day we indeed are allowed to feed on honey-dew and drink the milk of Paradise.

Creativity and the Brain

Our goal here is to drill down to the deepest level possible and attempt to find the neural basis of creativity. To reach that destination, we must take a few side excursions and learn more about the neural basis of normal thought. Attempting to understand how the brain thinks —an exceedingly complex topic—takes us near the edge of a modern scientific precipice. As we shall see, the human brain is certainly the most complicated organ in the body, and it may be the most complicated device on earth, or perhaps even in the universe. Only during the past ten to twenty years has it begun to give up some of the secrets about its near-miraculous activities and abilities. As we try to understand how the brain thinks, we will necessarily rely on a combination of established empirical facts and less-established theoretical speculations.

How Does the Brain Think?

Let me begin with a personal anecdote, which illustrates that an interest in understanding the neural basis of human thought, let alone the neural basis of creativity, is a recent phenomenon in the history of neuroscience.

Slightly more than thirty years ago, I was a medical student learning neuroanatomy. Neuroanatomy was considered by most med students to be the most dismal course we had to take. It consisted primarily of walking into a laboratory and looking at sliced slabs of formalin-fixed (and therefore very smelly) brain tissue, which had specific brain subregions marked on them. Our task was to learn all the different (and more or less meaningless) Latin names for the various regions and to attempt to reconstruct them three-dimensionally in our minds: names like the superior colliculus, the anterior fasciculus, the locus caeruleus, the substantia nigra, and even the substantia innominata. In fact, I don't think the instructors even cared much about three-dimensional reconstructions. Their biggest concern was how many of the picky little details we could remember. Our examinations consisted of walking around the laboratory room from one station to another, looking at the slab of brain tissue at any given station that had a pin in a particular region, and regurgitating its apparently meaningless name from memory.

Unlike most people in my class, I had decided to go to medical school because I wanted to understand how the human brain worked. How do we store memories? How do we generate spontaneous speech? How do we create music or literature? In summary, how does the brain think? I naively assumed that achieving this goal would be one of the purposes of studying neuroanatomy. However, this was a few years before the Society for Neuroscience was founded and also well before the field of cognitive neuroscience had begun to evolve. Of course, I was not aware of that small wrinkle in the history of neuroscience. In short, I was very ignorant.

I also assumed that medical school professors were, if not omniscient, then at least very wise. As the semester began to draw to a close, and I began to realize that none of my questions were being answered in the formal classroom teaching, I decided to begin to ask them. So, one day, I stuck up my hand during a lecture session and asked: "There is something important that we have never discussed in this class. How does the brain think?" The professor looked at me as if I had flown in from outer space, repeated my question in a mocking tone, and made facial and body gestures indicating that I had asked the most stupid question in the world. Many members of our nearly all-male class guffawed in sympathy with his reaction. I was of course mortified, although I did not comprehend why the question was considered to be so foolish. Only later did I realize that his response was one of self-defense. I had asked a question to which the omniscient professor could not even approximate an answer.

"How does the brain think?" remains a difficult and elusive question even now. But we are much closer now to being able to come up with answers. If we are embarking on a quest for Xanadu to understand the mystery of human creativity, we must begin by attempting to understand the complex workings of the human brain and the miraculous process of human thought.

A Primer of Brain Anatomy

First, you must learn a tiny amount of neuroanatomy, so that we can communicate through a shared vocabulary. Making the terminology of neuroanatomy interesting is still a difficult endeavor, but I can at least guarantee that the following description will be less boring than what I endured as a freshman medical student.

The human brain is composed of two major components. One is the cerebrum, which is divided into left and right hemispheres. The other component is the cerebellum (literally, "little brain"), located just below the cerebrum. Each of the cerebral hemispheres is divided into four lobes: frontal, temporal, parietal, and occipital. Oversimplifying somewhat, we can say that each of these lobes performs some specialized regional functions. The occipital lobes are used for vision, the

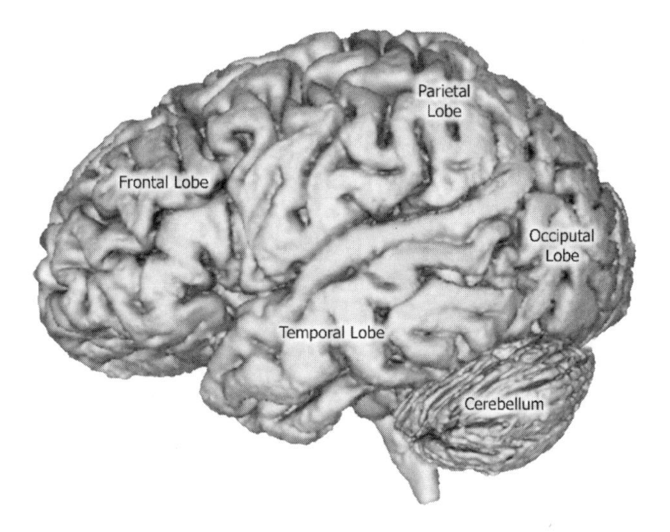

Figure 3-1. The brain, seen from the side, showing its four lobes and the cerebellum. This image of the brain was obtained from a living human being using Magnetic Resonance (MR) scanning technology. The multiple individual slices that are collected with an MR scan have been reconstructed using computer software developed by my imaging program at Iowa, known as BRAINS2.

temporal lobes for auditory perception and language, the parietal lobes for spatial perception and language, and the frontal lobes for "executive functions" such as abstract thinking, planning, and some types of memory. The lobes also contain regions that are less specialized, known as association cortex, which provide a place where several of the more specific functions can be linked together at the same time (i.e., "associated").

The surface of these lobes contains layers of nerve cells, known as neurons. Neurons are the basic unit for communication in the brain. The neurons are lined up in six layers throughout most of the cerebrum. This six-layered collection of nerve cells on the outer surface of the brain looks darker than the rest of the brain in brain slices, and is referred to as gray matter, or the cerebral cortex (Latin *cortex,* bark of a tree). The six-layered organization of the cortex is also sometimes referred to as "cortical columns," "modules," or "laminar units."

A neuron consists of a large cell body, which expands its activity and influence by several different kinds of extensions. Each neuron enlarges its capacity for sending and receiving information by extending branches from its surface, called dendrites (Greek *dendrites,* tree branches). Each of the dendrites further expands its informational capacity by having small nodes known as spines. On the spines are located synapses (Greek *synaptein,* to fasten together). The synapses are the discrete points that nerve cells use to connect to one another to send information back and forth. Connections between both nearby and distant brain regions are very important for "brainwork." These are also provided by extensions of the neurons that are known as axons, which function as "insulated wiring" between the cells, and which end in multiple terminals containing multiple synapses. Because the axons appear light in hue on sliced sections, these connecting fibers are generically referred to as white matter.

The neurons within the gray matter not only talk back and forth with their neighbors, but they also talk to nerve cells in more distant

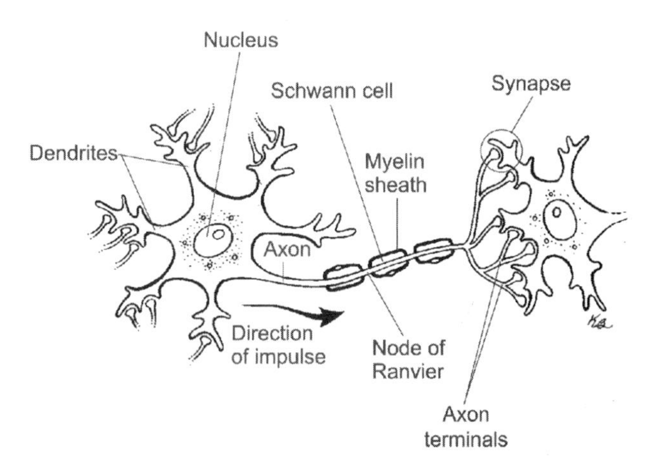

Figure 3-2. This is a simplified image of the structure of a neuron. It consists of a cell body containing a nucleus. The cell body is surrounded by dendrites, which expand the cell's communication capacities through having spines. The cell body also communicates through more distant neurons by sending out an axon. The axon is housed in a fat-rich myelin sheath, which provides insulation, and it sends out terminals. The final unit of communication is the synapse.

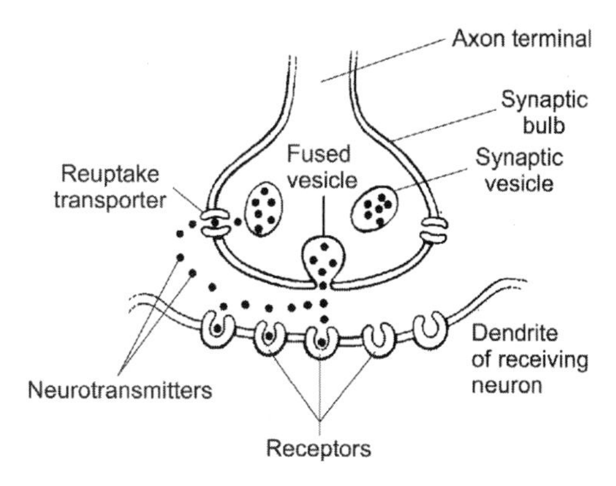

Figure 3-3. The synapse is the workhorse of the brain. It provides the mechanism by which neurons communicate with each other, by releasing neurotransmitters that are stored in synaptic vesicles that are located at the end of dendrites and spines. The neurotransmitters cross over to the cell bodies, dendrites, and spines of other neurons and occupy their receptors, causing them to become active.

regions—perhaps a few centimeters over in another gyrus, or perhaps many centimeters away in another lobe of the brain. For example, when our eyes record a visual image, this image is transferred back to the occipital cortex, where the specific features are perceived. That perception is then passed forward to another part of the brain specialized for interpreting the features. Where the percept goes depends on its nature and the context in which it is perceived.

For example, if the percept is an object such as a parked car or the face of another person, the information about the percept is transferred forward from the visual cortex in the occipital lobes to visual association cortex (the fusiform gyrus) and on to the temporal lobes. Its meaning is interpreted based on information about objects that is stored there. The interaction between the visual cortex and the other regions permits us to recognize that the car is a 911 Porsche or the face is that of Uncle Joe. This brain network has been called the *what* pathway, because it permits us to identify *what* we are seeing. If the percept occurs while we are in a car driving down a road, however, and our eyes and visual cortex transmit a different form of image, such as a car coming toward us from the opposite direction, then the information will be forwarded to another part of the brain, the parietal cortex, specialized for recognizing *where* an object is. This is the *where* pathway.

How do our brains spontaneously know which operation to perform, depending on the nature of the percept and the context that we are in? That is one of the miracles of ordinary human thought! This simple example of seeing a moving automobile illustrates the complexity of the thinking processes that our brains must perform, and do per-

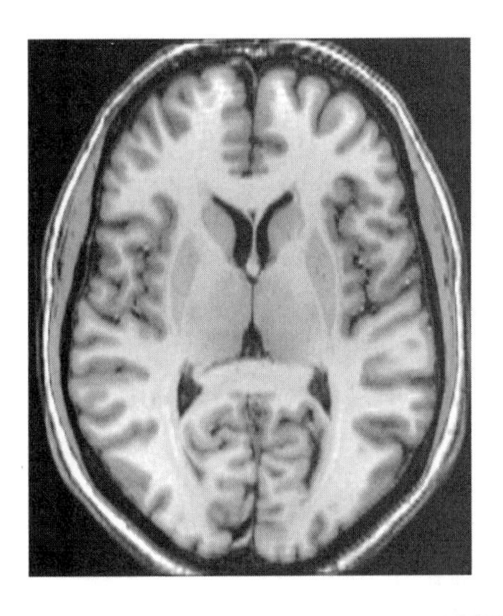

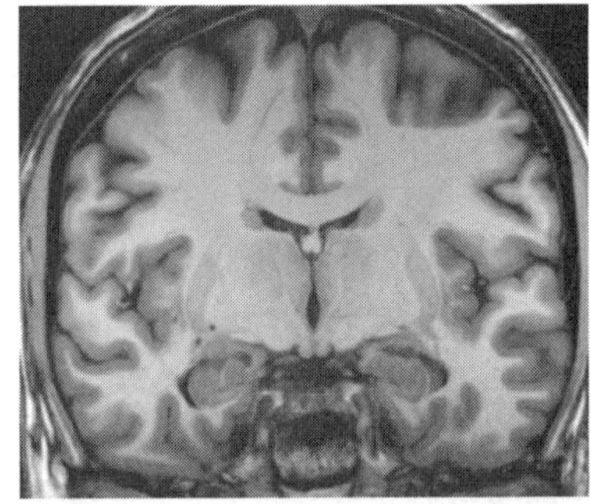

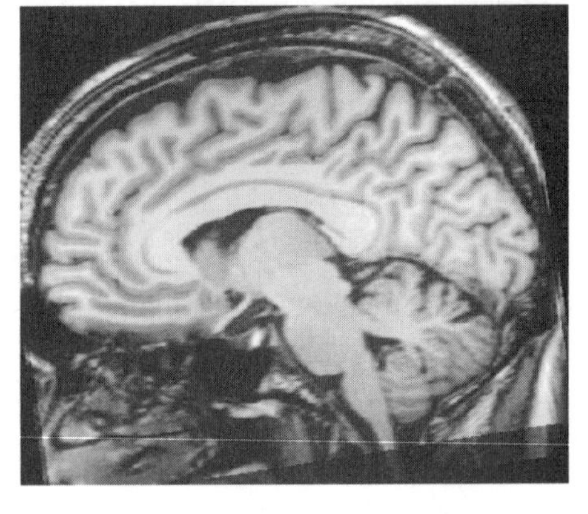

Figure 3-4. MR scan of a living human brain, showing how it looks when sliced in three different planes. The axial slice (upper left) runs in a plane parallel to the ground; the coronal slice (upper right) runs perpendicular to the ground; and the sagittal plane (lower right) runs from left to right. In each of these planes the cortical gray matter, or cerebral cortex is visible, running around the outside and in gyri inside the brain. The white matter is also clearly evident.

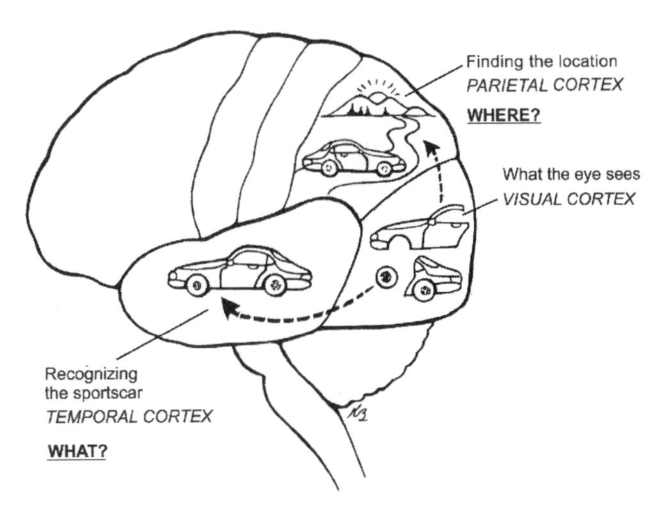

Figure 3-5. The where/what pathways of the brain

form rapidly, successfully, and easily throughout every millisecond of our lives.

The Complexity of Brain Networks

We are able to perform these complex mental activities in an effortless manner because of the extraordinarily complex organization of the human brain. The illustrations I have included show the structure of neurons, dendrites, spines, and synapses. What they do not portray is the way that these structures are linked to one another. The complexity of brain networks is so great that it is difficult to convey in words.

To begin with, the cerebral cortex is estimated to contain approximately 10^{11} (or 100 billion) neurons, while the cortex of the tiny cerebellum contains even more—approximately 10^{12} (or a trillion). In addition, we have islands of gray matter located in strategic positions within the sea of white matter fibers. These form relay stations between cortical regions and one another and are referred to as subcortical (under the cortex) regions or nuclei. Some examples include the thalamus, caudate, and putamen. Estimates of the number of cortical neurons do

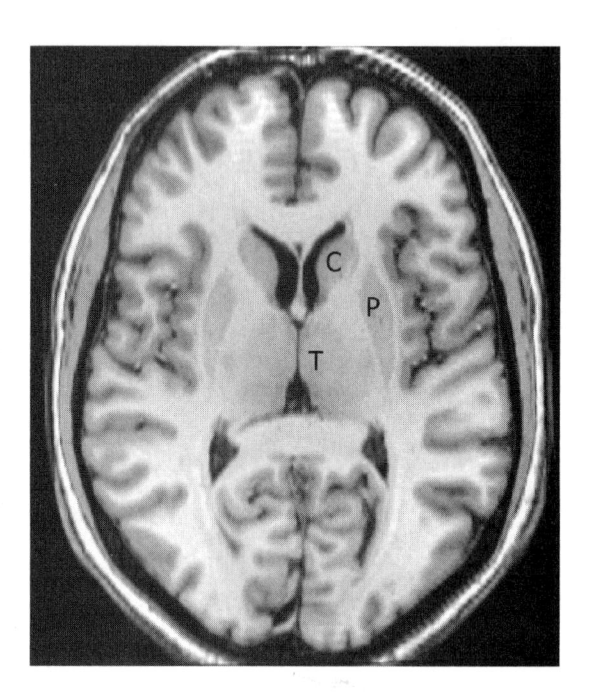

Figure 3-6. Subcortical structures

not include the neurons in these subcortical regions, which probably add a substantial amount (another few billion at the least), in addition to increasing the organizational complexity of the brain. So overall our brains contain more than a trillion neurons.

However, enumerating the number of neurons in the brain only tells a tiny part of the story. Each of these neurons is designed to make multiple connections to other neurons. The nerve cells multiply their connective capabilities by sending out dendrites, which in turn expand by adding spines. Along the spines are multiple synapses. At the axonal end of the neuron there are also axon terminals containing synapses. The synapses are the real "action sites" within the brain. There are different types of neurons, as defined by their number of axons and the complexity of their dendrites, and we do not have an accurate way to estimate the total number of synapses in the entire human brain. A typical estimate is that each nerve cell possesses approximately 1,000 to 10,000 synapses. If we multiply that times the number of neurons, we reach a number that is almost incomprehensible—somewhere in the neighborhood of 10^{15} or 10^{16}. After having heard so many people use the terms "zillion" or "gazillion," I decided to find out their mathematical meaning. It turns out that they have none. They just mean "big numbers." The number 10^{15} is technically known as quadrillion. Since we don't really know the number for sure, and since I like the way it sounds, I'm going to settle for quadrillion as the number of synapses in the whole brain. But suffice it to say that the number of connections in our brains is mind-boggling.

Having said that, I must add that discussing these numbers still makes the brain seem misleadingly simple. Adding to the complexity is the fact that the synapses connect cells to one another. How these connections occur is part of the complicated process of brain development and maturation. As our brains form during fetal life, nerve cells grow and establish connections to one another. Some of them are hardwired and genetically determined, but many are shaped by our experiences. Each neuron does its work by "talking" across synapses to multiple other neurons, often at more or less the same time, and each of these neurons is also talking to many others. (The technical term for these interacting neurons is *neural circuits.*) If we imagined this process as actual conversation, the cacophony would be infinitely louder than the loudest rock concert you've ever heard. The neural circuits of the brain are designed to monitor and modulate one another. Sometimes the connections send excitatory signals, and sometimes they send negative, or inhibitory, signals. Some connections create short feedback loops between neurons, and some have long loops that spread across longer spans of the brain. It is estimated that a large feedback loop covering the entire brain takes only five or six synapses. So the system is not only very complicated, but also very efficient.

Further adding to the awesome complexity of human brain networks is the fact that they are in a state of constant change. Our brains are living organs. They are constantly active, even when we are resting or sleeping. They use glucose (sugar) as their sole fuel, and they burn an average of 20 percent of the calories we use per day, while constituting much less than 20 percent of overall body mass. This is because our trillion neurons and quadrillion synapses are not static entities. They are living things, constantly busy, constantly making dynamic changes in response to external stimuli and internal states. We have words for some of these changes at a very abstract level, such as learning, remembering, or paying attention. But we do not have enough words to describe the nature, extent, or magnitude of these changes at the level of synapses or molecules, in part because we have not yet fully determined all of them. We have discovered some, such as a process known as long-term potentiation (the mechanism by which memories are formed in synapses). More will be illuminated during the coming years.

All thoughtful neuroscientists ponder the fundamental question about how the brain orchestrates these continuously occurring changes. What decides on which changes should occur? Is there a site within the brain that functions as an executive? A site that decides, for example, that we should memorize someone's phone number, rather than just forgetting it, as we do with most that we hear? A site that decides that we should glance down at our speedometer after seeing a highway patrolman and almost simultaneously put on the brakes? Such a site might be the essence of our self, something like an ego, a soul, or a source of consciousness. Some sites, such as the thalamus, have been proposed as serving such a role—most famously by the Nobel Prize–winning biologist Francis Crick and Christof Koch, his collaborator during his later years.

We really do not know the answer to this fundamental question. Right now we only have hunches, more pompously referred to in science as theories or hypotheses.

The hunch that is most widely accepted at present is that there is not a single site in the brain that acts as its executive. The "single site" hypothesis has a great appeal because of its simplicity. But it seems unlikely. Among other drawbacks, it raises yet another thorny question: how would the executive in the single site know which decisions to make? In a manner reminiscent of ancient religious and philosophical debates, we need to ask: if the thalamus (for example) is the "cause" of our brain's decisions about how to change, is there another "cause" that in turn directs the thalamus? That is, how does the thalamus know what to decide? What might that "cause" be? And is there yet another decision-maker ("cause") behind that one? The chain of causality inevitably becomes endless and unsatisfying, much as in medieval religious debates about the nature of the "final cause."

The Human Brain as a Self-organizing System

An alternative explanation is that the whole brain—or at least multiple key parts—participates in such decisions and acts as its own executive, based on events and information encoded in its reservoir from its past. It is what is known as a "self-organizing system," or SOS.

The concept of self-organizing systems is currently theoretically popular (especially in mathematics, physics, and computer science). Although it seems relatively novel, it is actually more than fifty years old. It derives from an earlier field, known as cybernetics, developed primarily by Norbert Weiner. Cybernetics investigated how systems were controlled by feedback mechanisms. The concept of self-organization was first proposed by an expert in cybernetics, W. Ross Ashby, in a book called *Design for a Brain,* published in 1952.

Current explanations of self-organizing systems tend to draw heavily on chaos theory, because they view the self-organization process as dynamic and nonlinear. The term "dynamic" means that the system is subject to frequent change, or is not "in equilibrium." The term "nonlinear" is a bit more complicated to explain. It is contrasted with the term "linear." A linear process is orderly, predictable, and relatively simple. If graphed, it forms a straight line. For example, if you hit a tennis ball with five pounds of force, it will travel with a certain predictable velocity. If you hit it with ten pounds of force, the velocity will double, and if you hit it with fifteen pounds, the velocity will triple. In mathematics, linear equations have a single solution. Nonlinear processes, on the other hand, are much more complicated. In linear systems, the effect is proportional to the cause, but in nonlinear systems the effect is less predictable. Small causes can have large effects. (An example would be the "butterfly effect" in chaos theory, whereby a butterfly flapping its wings in China can affect weather in New England.) And large causes can also have small effects. In mathematics nonlinear equations can have multiple different solutions. If graphed, nonlinear functions are described by curved lines and can be very complex. One reason for the complexity is that they often have to take interactions into account, especially the feedback from one component to another. The brain, as we have seen, is a mass of feedback loops.

A self-organizing system is one that is literally a whole that is greater than the sum of its parts. It is defined as a system that is created from components that are in existence and that spontaneously reorganize themselves to create something new, without the influence of any external force or executive plan. Control over a self-organizing system is not centralized. It is distributed over the entire system. A very simple example is how an audience may applaud after listening to a superb concert performance. First the applause is scattered randomly, with hands clapping in many different patterns. As the enthusiasm of the audience builds, a few people start clapping to a set rhythm. More slowly join in. Eventually the entire audience starts clapping in unison in an ordered rhythm, without any "executive" instructing them to do so. They have spontaneously organized themselves into a new and effective unit that expresses their appreciation for the concert that they have heard.

There are numerous examples of self-organizing systems, both small and large, from many branches of science. One is the formation of weather systems. Others from biology include the swarming of bees, the schooling of fish, or the flocking of birds. Ants appear to organize themselves spontaneously into functioning anthill colonies. Ecosystems modify themselves as weather changes, or as new predators or bacteria arrive. Snowflake crystals form into unique patterns. The economic market self-organizes in response to changing geopolitics and the waxing and waning of material resources.

The human brain is perhaps the most superb example of a self-organizing system that one can find. It is constantly and spontaneously generating new thoughts, often without any apparent external control. A simple self-organizing system can consist of as few as two components. The human brain, with its trillion neurons and quadrillion synapses, has nearly endless components to self-organize. Furthermore, because it is comprised of many large and small feedback loops, and because these can have both positive and negative input, it is the perfect organ for producing dynamic nonlinear thought.

What Is Human Thought?

Have you ever thought to ask yourself that question? Probably not. Because we think all the time, we are so familiar with it that we feel we understand it without any need to define or explain it. However, the next step in the journey to Xanadu must be to ask ourselves what human thought actually is, to come up with a plausible definition, and then explore how human thought arises from the human brain.

The brain engages in several different kinds of thought.

One is what we might call "ordered" or "conscious" thought. This is the thinking that we use when we perform specific mental tasks, such as producing spontaneous speech when answering a question or telling someone a story. In general, ordered, conscious thought has a temporal sequence that defines its order. It has a beginning, a middle, and an end. If the sequential order is broken, the thought no longer "makes sense." Although there are many other examples of ordered conscious thought, the generation of spontaneous language is a good place to begin a discussion of human thought because it is something that we already partially understand at the neural level.

The human capacity to generate an ordered sequence of words that "make sense" is a near-miraculous ability. It is a good example of "ordinary creativity," produced by a self-organizing system, for several reasons. First and foremost, most of the time that we speak, we are producing a sequence of words that we have not produced before—in fact, that no one has produced before. Unless we are saying our lines in a play or repeating clichés, we are producing language that is novel. We make up coherent sentences "on the fly," listening to ourselves speak while we are speaking, and planning what the next words will be as the words and sentences are produced. This is possible because the brain is a self-organizing system that can create novel linkages on a millisecond time scale. Much of this task involves the simultaneous generation of several components: a *discourse plan* that creates the overall shape of the language units to be spoken (usually a group of sentences connected by a unifying thought or theme); a *sentence plan* that formulates the individual sentences and pronounces them sequentially; and a search through the *verbal lexicon* for the appropriate words to place in order within the sentence. In addition, the brain also has to use its motor components to permit us to move our lips and tongue and palate so that the speech is well-articulated, as our auditory system listens to what is being said and prepares the other components to make modifications in the discourse plan, sentence plan, and verbal lexicon choices. Making matters more complicated, we often do this as we watch the face and body language of people listening to us, sometimes deciding to make changes in the discourse plan as we see them wince . . . or laugh or smile.

Based on the study of people with brain injuries (known in the neuroscience trade as "lesion studies"), we have had a reasonable map of the large-scale components of the language system for many years. This information has been supplemented and modified since the 1990s by new *in vivo* neuroimaging technologies that are used to map human brain function. These technologies, positron emission tomography (PET) and functional magnetic resonance imaging (fMRI), have permitted neuroscientists to visualize human brain activity in living people while they are conducting mental tasks.

We now know that the human brain contains a group of dedicated distributed nodes that we use to produce and understand spoken and written language. By "dedicated" I mean that a given node is relatively specialized to perform a given function. By "distributed" I mean that the nodes are widely dispersed through the whole brain and interact with one another across relatively long distances.

Broca's area was the first of the nodes in the language system to be

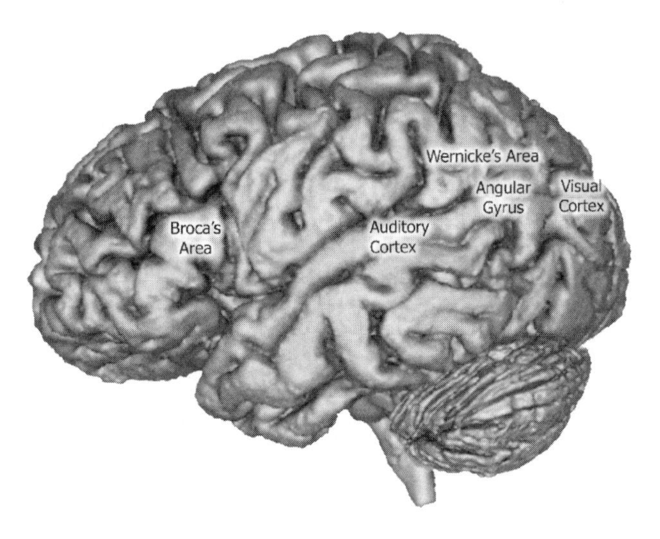

Figure 3-7. The nodes in the language network are widely distributed throughout the brain. The visual area at the back, in the occipital cortex, is the first stop when we read. The auditory cortex is the first stop when we listen to speech. Wernicke's area and the angular gyrus are association cortices. Broca's area was considered to be the site where speech is generated for many years, but we now know that it serves other functions as well; for example, it is highly developed in concert orchestra musicians, as described in chapter 6.

discovered—in the mid-nineteenth century by the French physician Paul Broca, after whom the area is named. He cared for a patient who had suffered a stroke and was no longer able to produce fluent grammatical speech, although he could understand what was said to him. After the patient's death in 1861, Broca examined the brain and determined that the stroke had principally affected a posterior inferior frontal region in the left hemisphere. He deduced that this region was responsible for language production. For more than a century, physicians and neuroscientists believed Broca's declaration that "we speak with our left hemisphere" and held firmly to the doctrine of hemispheric specialization. This doctrine proposes that the left hemisphere is primarily responsible for language functions, while the right hemisphere is specialized for nonverbal visual/spatial functions. This doctrine is now called into question, however, by *in vivo* neuroimaging studies, which indicate the presence of a "mirror Broca's area" in the right hemisphere that is also activated during language production. This may explain why some people are able to recover language function after a left hemisphere stroke.

The mapping of language nodes has continued on through the nineteenth and twentieth centuries. At first blush, we would be likely to think that our ability to communicate in language is a unitary ability. We would not assume that it is divided into the ability to speak fluently and grammatically, the ability to assemble the appropriately meaningful group of words as we speak, the ability to understand what others say, the ability to read, and the ability to write. But that is what was subsequently discovered, through a long series of lesion and imaging studies. Although we use one word, language, to refer to this set of abilities, language is comprised of multiple components, which are facilitated through the activity of different and widely distributed brain regions.

The German psychiatrist Karl Wernicke, who practiced in the late nineteenth century, observed that patients who had lesions in an inferior parietal region (now known as Wernicke's area) had significant problems with understanding what was said to them and with producing coherent speech. Although they could generate strings of words, the meaning was garbled and did not make sense. Other patients were subsequently observed who had an injury to a different parietal region, the angular gyrus, who could both speak and understand but could not read or write. Most of these cortical regions are known as "association" cortex. That is, their primary job is to perform linkages of associations. For example, Wernicke's area receives input from the auditory cortex, a more highly specialized part of the brain that hears sounds. Sounds received in the auditory cortex are just that—sounds. They are "understood" when they are connected with Wernicke's area, which recognizes the sounds as words that have specific meanings and that are linked in a specific order to form comprehensible sentences. The angular gyrus receives input from the primary visual cortex and recognizes groupings of letters as words. We presume, primarily on the basis of imaging rather than lesion studies, that the sounds or letters are also referenced against a verbal lexicon residing in the temporal association cortex, where the sounds and words are connected to their associated memories and given "meaning."

The associations that form meaning are known as "qualia" in the terminology of philosophy. The meanings, or qualia, that are associated with a word or concept may vary subtly with each given individual. For example, "boat" is likely to have different associations for a landlocked Midwesterner than for a coastal Bostonian. Among Bostonians, the word may vary in meaning depending on whether the person rows the Charles River, sails a sailboat, or owns a fishing boat, or all three. It is likely that one factor contributing to literary creativity is having a lexicon not only large in quantity of words but also rich in associated meanings for each word.

Unconscious Thought: The Edge of the Mind's Precipice

Producing sequentially ordered speech is a conscious activity that we intentionally perform, drawing on our brain's capacity to act as a self-organizing system. This is an excellent example of the human capacity for ordinary creativity, and it may also play a role in extraordinary creativity. However, we all also have an unconscious mental life, and that life may be highly relevant to extraordinary creativity as well. The introspective accounts by Mozart and others presented in chapter 2 highlight the importance of unconscious creative processes that we "run in background" during waking hours and that may also occur during sleep. We might think of unconscious mental activity as "disordered thought," because it does not tend to have the same linear pattern as ordered thought.

In general, neuroscientists interested in human cognition prefer to study its simpler components—things like visual perception, auditory perception, types of attention, or types of memory. They choose this direction because these simpler components are more easily studied using scientific methods. A few brave souls have ventured into "softer" and "less scientific" topics, such as emotion. Even fewer have attempted to study and explain consciousness, which many consider to be "the last frontier" of the brain. But these brave souls have included some of the best minds in neuroscience, such as Gerald Edelman, Francis Crick, and Christof Koch. Much else that has been written on this topic has been done by philosophers, such as John Searle and Daniel Dennett.

Almost no one has ventured to advance beyond that frontier and to risk taking a precipitous fall over the edge—into the realm of the unconscious.

The unconscious has, however, been a topic of great interest to psychiatrists for many years. Sigmund Freud in particular turned the world's attention to the study of the unconscious by proposing that unconscious processes might influence our thinking and behavior in ways of which we are not consciously aware. We have lived so comfortably (or uncomfortably) with Freud's ideas for the past century that it is sometimes difficult to appreciate the great originality of some of them. Many became rigidified and even stultified through the creation of multiple Psychoanalytic Institutes, which often warred with one another. Others became trivialized through jokes about cigars, couches, and other psychoanalytic paraphernalia. Some people say that Freud should have won a Nobel Prize, and others argue that it should have been for literature rather than medicine. There may be some truth in this latter point, but there is no Nobel Prize for the field in which he truly excelled: philosophy of mind. Perhaps his greatest contribution was his recognition of the existence of the unconscious and his discussion of its relationship to conscious processes. This novel idea expanded in many directions during the twentieth century, having an impact not only on psychiatry and neurology, but also on philosophy, literature, art, and music.

As a doctor who lived in the world of both psychiatry and neurology, Freud saw many patients—predominantly Viennese women—who suffered from a variety of unexplained and inexplicable physical complaints. Without any clear medical explanation, they would become speechless, blind, or paralyzed. Such unexplained symptoms have been recognized since classical times and called "hysteria." In classical times they were attributed to a displaced uterus (Greek *hysteros*) that was creating pressure on other body organs. Working in an era when the principles of thermodynamics were being studied in physics, Freud suspected that these women instead had disturbed psychodynamics. They might have repressed thoughts in their brains that were somehow putting pressure on other parts of their bodies. But how could he probe into their minds and brains? How could he relieve the pressure?

He developed the idea that the problem might be repressed memories, of which his patients were unaware, but which had collected in the "unconscious" compartment of their minds and were making physical mischief. He decided that if patients could recall these memories and release them by talking about them, they might relieve the symptoms. But how to get them to recall and release the memories? Perhaps it might happen if they would lie down on a couch, think freely, and say whatever came into their minds. This process was called *free association* because it relied on pulling up a variety of associative links that lurked in the brain at an unconscious level.

Initially, Freud performed this treatment when sitting behind the patient on the couch and putting gentle pressure on her forehead to help release the free associations. Victorian gentleman that he was, he was astonished at the things that began to come out of their mouths. Victorian ladies, after all, had to use the word "limb" to refer to the legs of a chair, since the real word was considered too graphic, and improper. But these patients actually talked about SEX! They described having it with their fathers and not having it with their husbands! Free association was off and running as a way of probing into mind and brain. And here we are a century later in the world of MTV and graphic sexual overtones in advertising, with many of us mourning the loss of any sexual restraint or romantic sexual mystery at all in the society around us.

The concept of free association did not in fact begin with Freud. It was created by early association psychologists such as Wilhelm Wundt. (Wundt was one of the nineteenth-century founders of modern psychology, which began to blossom in the twentieth century and continues today.) The meaning and applications of free association expanded, however, after Freud popularized it through psychoanalysis and psychodynamic theories. The term has developed a variety of meanings, depending on the discipline that refers to it. "Free association," as expanded from its roots in association psychology by pioneers in psychoanalysis, is a method for examining thinking processes that do not involve conscious organization of events into a temporally-linked account. To the extent that the thinking is uncensored, it taps into *primary process thinking,* that is, into unconscious thought that is primitive in organization and often in content.

Free association also carries less technical connotations outside the domain of psychoanalysis. In this more general sense, it is also a method for exploring human thinking that has had a powerful impact on artistic creativity, influencing writers to portray the "stream of consciousness" in literary form in novels such as James Joyce's *A Portrait of the Artist as a Young Man* and William Faulkner's *The Sound and the Fury.* The term is also used in "ordinary language" to refer to processes such as daydreaming, brainstorming, or letting the mind wander freely or think randomly. Free association is the kind of mental process that occurs when a person eliminates motor and sensory input by stretching out on a bed with eyes closed and "just thinks." This mental process, which connects apparently unlinked things without conscious effort, is an important resource for creativity.

For most serious neuroscientists, attempting to study the neural basis of this type of thinking is "over the edge." It is too difficult to examine experimentally and scientifically. Always an adventurer myself, however, I decided several years ago to take the plunge over the precipice and to study the neural basis of free association using neuroimaging technology to obtain measurements of cerebral blood flow to determine which regions of the brain became activated. This was the first (and to my knowledge, so far the only) effort to examine how and where the brain generates unconscious thoughts.

In order to explain how I did the study, I need to introduce another important concept about how the mind and brain work. The thinking tapped during free association (in its full breadth of definitions) draws on a subtype of memory known as *episodic memory*, which was originally defined by the great contemporary guru of memory, Endel Tulving. Episodic memory is autobiographical memory, the recollection of information that is linked to an individual's personal experiences. It is called "episodic" because it is composed of a series of events (episode = event in Greek), which are sequentially ordered in time. Its time-linked nature is a crucial component, and it may encompass both looking backward into the past or looking forward and "remembering the future." The capacity to place events in time and to reference them to oneself may form the basis for self-awareness or consciousness.

Episodic memory is sometimes contrasted with another memory system, *semantic memory*, which is impersonal and not as inherently time-linked. It comprises an individual's repository of general information. More generally, semantic memory includes a broad range of information about the world that is detached from personal experience and that is defined by a cognitive as opposed to autobiographical reference.

Episodic memory is probably uniquely human. Episodic memory is frequently used to generate coherent speech, as described earlier, as well as other conscious processes that require drawing on personal memories. Episodic memory permits the individual to reference his or her personal experiences in both time and space; as such it creates the experience of consciousness, the sense of individual identity, the ability to refer one's own experience to that of others, and the capacity for introspection. Most of the time that we use episodic memory, it is conscious and focused on some particular purpose, such as describing the day's activities when we sit with a spouse or friend and chat over dinner after work.

Episodic memory is also used for free association. The kind of episodic memory tapped during free association is more mysterious. It is less clearly sequential and time-linked. It may be the repository of information that is stored deeply and is therefore sometimes less consciously accessible. It draws on those freely wandering and undirected associative thoughts that constitute primary process thinking. It is a resource not only for the creative process, but also for meditational states, religious experiences, and dreams.

Several years ago I decided to probe the neural basis of these two types of episodic memory using the tools of neuroimaging—specifically positron emission tomography (PET), which is used to visualize and measure changes in regional brain blood flow as mental activity changes from one type to another. In their early days PET studies typically used an "experimental condition" and a "control condition." The experimental condition was the topic being explored. For example, the brain regions needed for verbal fluency might be probed by asking the subject to say all the words that she could think of that start with the letter "D." The control condition was typically lying with eyes closed and "resting."

Although neither a Freudian nor a psychoanalyst, I knew enough about human mental activity to quickly perceive what a foolish "control task" rest was. Most investigators made the convenient assumption that the brain would be blank or neutral during "rest." From introspection I knew that my own brain is often at its most active when I stretch out on a bed or sofa and close my eyes. That relaxed position induces the pleasurable experience of free-floating creative thought. But I also knew from interviewing psychiatric patients that their brains do not rest either. When they close their eyes and "rest," if they don't speak, they engage in silent free association. When they do speak, they free-associate, at least after they have become trained and comfortable with the process.

So my PET study examined a simple and specific episodic memory task. All subjects were asked to describe what they did that day, starting from when they got up in the morning. This generated a narrative comprised of conscious, logical speech that drew on episodic memory. I called this *focused episodic memory.* I then compared this state of mind with what my colleagues were calling "rest," but which I called (a bit tongue-in-cheek, to make my point) *random episodic silent thought,* which created the acronym REST. For me, this too was an experimental task, not an idle control state. But it was one that no one had ever recognized as such or studied. The "control task" for my study was instead bland, boring semantic memory. This research design would permit me to probe the brain regions used for conscious episodic memory as compared with unconscious episodic memory, or REST. Most importantly, it would give me a window into the neural basis of the unconscious.

Can you guess what I found? Which brain regions would you predict to be active during randomly wandering unconscious free association?

Not surprisingly, it was almost all association cortex. It was those areas in the frontal, parietal, and temporal lobes that are known to gather information from the senses and from elsewhere in the brain and link it all together—in potentially novel ways. These association regions are the last to mature in human beings. They continue to develop new connections up to around the early twenties. They are also much larger in human beings than in other higher primates. They have a more complicated columnar organization than other parts of the brain. They receive input from primary sensory or motor regions, from subcortical regions such as the thalamus, and from one another. Presumably this organization is used to permit the brain's owner to integrate the information he or she receives or possesses and to produce much of the activity that we refer to as "the unconscious mind." Apparently, when the brain/mind thinks in a free and unencumbered fashion, it uses its most human and complex parts.

For our purposes at present, focused episodic memory is less interesting, but the story would be incomplete if I left it out altogether. Several regions were active in focused episodic memory that were not involved during REST. They are primarily areas in the temporal lobe that are used when we recall things, a process called memory retrieval. Portions of the language circuit, described earlier, were also active because the subjects were speaking. Two regions were activated during focused episodic memory that were also active during REST: association regions in inferior frontal lobes, and a parietal association region known as the precuneus. The psychological commonality between the two tasks is that both involve something personal and highly individual, whether it be a past experience or unexpressed private silent thoughts. This inferior frontal region is known from lesion studies to be involved in social awareness and the ability to possess a value system and experience guilt; lesions in this region tend to produce uncensored and potentially antisocial behavior. The precuneus has received little study, since it is rarely damaged in strokes or accidents, but its location and structure suggest it to be major association cortex that could subserve a variety of cognitive functions, including memory functions. Perhaps the connectivity between the medial inferior frontal region and the precuneus represents a network through which personal identity and past personal experiences are interlinked, with the net interactions permitting us to move between self-awareness and disengagement, censorship and freedom, or consciousness and the unconscious.

The Neural Basis of Extraordinary Creativity

It is obvious that there is a neural basis for ordinary creativity. As human beings, all of us create new language every time we speak, using the multiple nodes in our language circuits. We all make connections between various words and ideas using our association cortex. We can perform tasks that require focused episodic memory, such as recounting personal experiences. We all have brains that are self-organizing systems. We are all able to think in nonlinear dynamic ways. But are these the same properties that produce extraordinary creativity as well? Does the extraordinarily creative person simply have a mind/brain that differs only in the amount or extent of these properties? Or does that person think in a truly different way? And if different, how so?

At this stage of our knowledge, we only have hunches, buttressed by modest evidence.

Very possibly, some successful creative people are running on neural processes that are an enlarged or enriched variant of ordinary creativity, at least during some of the time that they are creating. Think, for example, of the process of writing a novel. Writing a novel takes a relatively long time, at least a year and sometimes more. Most novelists have very regular habits. They typically get up and do their writing during the morning and early afternoon. They write even when they feel they have nothing to say. (They have learned that otherwise they might not write at all.)

And much novel-writing is akin to consciously generating spontaneous speech. The short-term discourse plan is there, and the sentences are created and the words are chosen "on the fly," but simply at a slower pace because the motor task of writing is slower than the task of speaking. Much of this is in the domain of ordinary creativity. Most novelists have larger vocabularies than average. They have the opportunity to hone their skills through practice. When not writing, they often carry notebooks, in which they jot down interesting observations to use at a later time, perhaps when they get stuck on defining a character or finding the best twist for the plot.

But then there are moments . . .

Sometimes a new character emerges out of nowhere and becomes the nidus for an entire novel. Sometimes a wholly new idea occurs to a writer, as when Joyce decided to write in stream-of-consciousness or Tolkien conceived of hobbits and Middle Earth. Sometimes the prose just comes pouring forth from brain to fingers to keyboard to page without any conscious effort at all. Those are times when extraordinary creativity is the neural mode. Generally, this "extraordinary creativity mode" cannot be sustained throughout the entire long period of novel writing, but it may comprise anything from a small to a significant part of it.

Furthermore, in thinking about whether extraordinary creativity and ordinary creativity are qualitatively different, we cannot ignore the introspective descriptions we have from figures such as Coleridge, Poincaré, and Simon. They have described mental activities that are anything but ordinary—extraordinary examples of what must be called extraordinary creativity. It seems that extraordinary creativity is characterized by thought processes that are different from the ordinary, and presumably by neural processes that differ qualitatively as well as quantitatively. Let us consider some of these examples again.

Coleridge says:

The Author continued for about three hours in a profound sleep, at least of the external senses, during which time he has the most vivid confidence, that he could not have composed less than from 200–300 lines; if that indeed can be called composition in which all the images rose up before him as *things*, with a parallel production of the correspondent expressions, without any sensation or consciousness of effort.

Neil Simon says:

I slip into a state that is apart from reality.
I don't write consciously—it is as if the muse sits on my shoulder.

Mozart says:

When I am, as it were, completely in myself, entirely alone, and of good cheer—say traveling in a carriage, or walking after a good meal, or during the night when I cannot sleep; it is on such occasions that my ideas flow best and most abundantly. *Whence* and *how* they come, I know not; nor can I force them. . . . my subject enlarges itself, becomes methodized and defined, and in the whole, though it be long, stands almost complete and finished in my mind, so that I can survey it, like a fine picture or a beautiful statue, at a glance. Nor do I hear in my imagination the parts *successively*, but I hear them, as it were, all at once *(gleich alles zusammen)*. What a delight this is I cannot tell! All this inventing, this producing, takes place in a pleasing lively dream.

Tchaikovsky says:

Generally speaking, the germ of a future composition comes suddenly and unexpectedly. If the soul is ready—that is to say, if the disposition for work is there—it takes root with extraordinary force and rapidity, shoots up through the earth, puts forth branches, leaves, and, finally, blossoms. I cannot define the creative process in any other way than by this simile.

Poincaré says:

One evening, contrary to my custom, I drank black coffee and could not sleep. Ideas rose in crowds; I felt them collide until pairs interlocked, so to speak, making a stable combination. By the next morning I had established the existence of a class of Fuchsian functions, those which come out from the hypergeometric series; I had only to write out the results, which took but a few hours.

Spender says:

Inspiration is the beginning of a poem, and it is also its final goal. It is the first idea which drops into the poet's mind and it is the final idea which he at last achieves in words.

In many cases these individuals are describing a process during which a composition or an insight occurs in a flash. Incredibly, sometimes the entire creation is all there in the mind/brain and needs only to be written down. The process seems to be inspired. It sometimes occurs when in a dreamlike state, or even during sleep. This appears to be a different process from the conscious production of a stream of speech or writing. It happens far more rapidly and produces something substantial that is completely new.

One may conclude not only that extraordinary creativity is at least sometimes based on a qualitatively different neural process than ordinary creativity, but that it at least sometimes arises from that "over the precipice" component of human thought that we call the unconscious. What is that process, and how does it arise?

We have seen from the study of REST, which tapped into the unconscious, that free-floating and uncensored thought (primary-process thought, primitive thought, original thought) occurs when multiple regions of our highly developed human association cortex interact with one another. When this occurs, the brain is working as a self-organizing system, but in a different way. Think of Poincaré's impression that "ideas rose in crowds; I felt them collide until pairs interlocked, so to speak, making a stable combination."

These introspective accounts are describing a process during which thought is not only nonsequential or nonlinear, but during which nonrational unconscious processes play a role. It is as if the multiple association cortices are communicating back and forth, not in order to integrate associations with sensory or motor input as is often the case, but simply in response to one another. The associations are occurring freely. They are running unchecked, not subject to any of the reality principles that normally govern them. Initially these associations may seem meaningless or unconnected. I would hypothesize that during the creative process the brain begins by *disorganizing,* making links between shadowy forms of objects or symbols or words or remembered experiences that have not previously been linked. Out of this disorganization, self-organization eventually emerges and takes over in the brain. The result is a completely new and original thing: a mathematical function, a symphony, or a poem.

If this hypothesis is correct, extraordinary creativity *is* qualitatively different from ordinary creativity. The underlying neural processes are distinct. They proceed by tapping into the unconscious in ways that possessors of ordinary creativity alone are usually unable to do. They begin with a process during which associative links run wild, creating new connections, many of which might seem strange or implausible. This disorganized mental state may persist for many hours, while words, images, and ideas collide. Eventually order emerges, and with it the creative product. Possessors of extraordinary creativity are apparently blessed with brains that are more facile at creating free associations. Neurally they may have enriched connectivity between their various association cortices, or they may even have different kinds of connectivity. Put simply, they are gifted with unusual brains that permit them to see and think in ways that are not accessible to ordinary mortals. This capacity is both a blessing and a curse, for it makes the creative person not only creative but also vulnerable.

4

GENIUS AND INSANITY

Creativity and Brain Disease

Those who have become eminent in philosophy, politics, poetry, and the arts have all had tendencies toward melancholia.

—Aristotle, *Problemata*

The lunatic, the lover, and the poet,
Are of imagination all compact.

—William Shakespeare, *A Mid-Summer Night's Dream*

Great Wits are sure to Madness near ally'd:
And thin Partitions do their Bounds divide.

—John Dryden, *Absalom and Achitophel*

"How could you believe you were recruited by aliens from outer space to save the world?"
"Because the ideas I have about supernatural beings came to me the same way that my mathematical ideas did, so I took them seriously."

—Sylvia Nasar, *A Beautiful Mind*

Recent books and films such as *Shine* and *A Beautiful Mind* have highlighted the fact that very gifted people may suffer from mental illness—or the converse, that mentally ill people may be highly gifted. David Helfgott and John Nash both displayed significant levels of creativity, although in very different areas. One was a talented musician, while the other was an inventive mathematician. Both probably suffered from schizophrenia, although Helfgott may have been at the affective end of the schizophrenia spectrum (known as schizoaffective disorder). What is the relationship between their mental illness and their creativity? Did mental illness facilitate their unique abilities, whether it be to play a concerto or to perceive a novel mathematical relationship? Or did mental illness impair their creativity after its initial meteoric burst in their twenties? Or is the relationship more complex than a simple one of cause and effect, in either direction?

John Nash and David Helfgott are not alone.

Many other gifted people have also suffered from serious mental illness, and their careers also raise similar questions.

Vincent van Gogh, for example, died at thirty-seven by suicide, having suffered during his last years from intermittent episodes of mania and depression, and having completed more than three hundred of his best paintings during the last year and a half of his life. His singular style ushered in an era of experimentation loosely known as "modern art." In another time, and while pursuing a very different goal, Martin Luther experienced terrible periods of depression, but also periods of manic energy and prodigious productivity. His despair over not being able to satisfy the demands of his conscience led him to proclaim the importance of "faith" rather than "works," launching the movement that came to be known as the Reformation. After writing his Ninety-five Theses and nailing them to the church door in Wittenberg in 1517, thereby opening a floodgate of religious change, he turned his enormous energies to writing theological tracts to support his position. These tracts, written in vernacular German in an accessible and vigorous style, argued the importance of basing religious beliefs and practices on direct interpretation of the Bible rather than on Church authority. In effect, he made "every man his own priest." Protestantism, the religion of those who protested the long-established and uncontested authority of the Roman Catholic Church, was launched through his combination of introspective self-criticism and exuberant energy. His alternations of mood changed the course of history.

It is not difficult to create a long list of highly gifted or creative people who suffered from mental illness. They come from all domains, including music, art, dance, poetry, drama, fiction, physics, mathematics, biology, philosophy, and politics. The list of names includes people such as John Nash, Isaac Newton, Friedrich Nietzsche, Leo Tolstoy, Virginia Woolf, Samuel Johnson, Jonathan Swift, Ernest Hemingway, Abraham Lincoln, Theodore Roosevelt, Oliver Cromwell, John Stuart Mill, Robert Schumann, Gaetano Donizetti, Ludwig von Beethoven, Robert Lowell, Graham Greene, John Berryman, Anne Sexton, Vaslav Nijinsky, Joan Miró, and many more who are eminent and admired.

Early Explorations of Genius and Insanity: The Anecdotal Era

Although an association between "genius and insanity" was noted in classical times and continued to be observed through succeeding eras, systematic investigations of the topic did not begin until the nineteenth century. One of the earliest intensive discussions was written by Cesare Lombroso, an Italian psychiatrist who was also interested in criminality. Lombroso believed that criminal behavior was inherited and not triggered in any way by social and environmental influences such as poverty, poor parenting, or lack of education. He devoted a great deal of his career to identifying the "physical markers" of the criminal person, such as prominent lips and enlargement of the nose. Much of this work seems naive and out-of-date today. But Lombroso's interest in the heritability of mental and social traits also helped launch explorations into the nature of creativity. His book *L'Uomo di genio* (1864) attracted international interest. It appeared in English, as the *The Man of Genius,* in 1891. Lombroso believed that both insanity and creativity were hereditary, and that they often tended to occur within the same person.

Francis Galton was the next to weigh in on the topic. He is a fascinating figure, whose own life illustrates both the familiality of creativity and its multifaceted nature. He was a cousin of Charles Darwin. His mother was the daughter of Erasmus Darwin, a prominent eighteenth-century physician who was also a poet and botanist. Thus Erasmus Darwin was grandfather to both Francis Galton and Charles Darwin, two of the most eminent scientists of the nineteenth century. Interestingly, although both came from highly educated families and both considered medicine as a career, neither practiced. Apart from medicine, neither had much formal training in science or mathematics. Both were essentially "gentlemen amateurs" who were self-taught scientists. They created new ideas simply by using their own observational skills and imaginative insights.

As a child, Galton appeared to be a prodigy. He was educated by his doting older sister Adele for his first five years. Just prior to his fifth birthday, he wrote her the following letter (reproduced here from Terman's 1917 study):

MY DEAR ADELE,

I am 4 years old, and I can read any English book. I can say all the Latin Substantives and Adjectives and active verbs besides 52 lines of Latin poetry. I can cast up any sum in addition and can multiply by 2, 3, 4, 5, 6, 7, 8, [9], 10, [11]. I can also say the pence table. I read French a little and I know the clock.

FRANCIS GALTON
Febuary 15, 1827

The numbers 9 and 11 are in brackets because little Francis apparently felt he may have exaggerated his multiplication skills, and so he pasted a piece of paper over one and scratched the other out with a knife. Note that the spelling is perfect, apart from "Febuary."

He studied medicine for several years, but never practiced. He also began a mathematics course at Cambridge, but was unable to complete it due to a "breakdown." His recent biography does not provide any information about the nature of the breakdown, and so we do not know whether his interest in hereditary genius was related to that experience. He traveled widely in Africa and wrote books describing his observations. After his cousin Charles published *The Origin of Species* (1859), which suggested that physical characteristics were both heritable and prone to evolve, Galton became interested in the possibility that intellectual traits might also be heritable and subject to evolutionary change. These ideas reached their culmination in his book *Hereditary Genius* (1869).

Galton made many discoveries that provided a foundation for later science. Perhaps his greatest contribution was his development of the concepts of the standard deviation and the correlation coefficient. He developed scatter plots (diagrams that show scattered points of individual data using an *x* and *y* axis to visually portray the relationship between two things that have been measured). He used the scatter plots to examine the relationship between measures such as head size and indicators of intelligence. He then recognized that scatter plots could be represented numerically as correlation coefficients (indicators of how closely the measurements are related). For example, if head size and an indicator of intelligence (such as good grades in school) are closely related, they would be tightly clustered in a neat line on the scatter plot, and the calculation of the correlation coefficient would produce a relatively large number, such as 0.8. (A perfect correlation is calculated as 1.0, but this essentially never occurs.) If they were not closely related, the scatter plot would look diffuse, and the correlation coefficient would be small (e.g., 0.2). This contribution, further developed by the great statistician Karl Pearson, led to the creation of a major field in modern statistics.

Galton was also among the first to popularize the interrelationship of genetic and environmental factors as "nature versus nurture," a phrase that still pervades modern discussions on the topic. As he attempted to explore scientifically the interactions between genetic and environmental influences, he was the first to develop the method of studying identical and nonidentical twins in order to describe the relative contributions of genetic and nongenetic influences. And he recognized that fingerprints are just that—individual fingerprints—and introduced them as a method for unique identification of individuals. This finding was soon adopted by Britain's Scotland Yard. In short, Galton was a remarkably creative man who made numerous original scientific contributions.

Francis Galton's *Hereditary Genius* was published in 1869, shortly after the publication of Lombroso's work on "the man of genius." Galton used a systematic and objective approach to reach his conclusions. However, his findings were heavily biased by his belief in eugenics and his desire to show that genius was hereditary. He chose cases that would prove his point. The wealth of pedigrees in *Hereditary Genius* nevertheless makes for fascinating reading. The book also provides a vivid snapshot of an era in which prejudices abounded and the concept of "political correctness" did not exist. Galton assumes, for example, that geniuses are male and that "negroes" are intellectually inferior. His arguments in favor of improving the human race by eugenics paved a wide road down which Nazi fascism and anti-Semitism could subsequently march, culminating in the horrors of the Holocaust. Much as Einstein did not anticipate the spectre of world devastation to which the atomic bomb would lead, it is unlikely that Galton foresaw the abuses to which his theories would be put.

Later Victorian and post-Victorian writers also compiled works on genius and insanity. Examples include Hirsch's *Genius and Degeneration* (1896), Nisbet's *The Insanity of Genius* (1900), and Hyslop's *The Great Abnormals* (1925). But unlike Galton, most of these authors made no effort to be scientific. They relied instead primarily on anecdotes. They began by identifying cases that illustrated an association between creativity and mental illness. They did not try to determine how common or uncommon the association was, and they certainly did not entertain the possibility that it might not exist at all (the hallowed "null hypothesis" of modern scientific method). They raised a series of interesting questions without answering them: Is there really more psychiatric illness among creative people than in the general population? If there is more, how common is it? Do creative people have a particular type of psychiatric illness, or do they have a wide range of types? If an association exists between psychiatric illness and creativity, is it with specific types of creativity (such as writing or painting), or does it occur in all types of creative people? If there is an association, does it tend to be familial? If the association is familial, is it genetic, nongenetic, or both?

Like Galton, Havelock Ellis was another student of creativity who was himself an interesting exemplar of giftedness and creativity. Slightly older than Galton, he also grew up during the prudish and mannered Victorian era. He eventually rebelled against this environment by developing a strong interest in the nature of sexuality and is best known for his contributions to that field. He decided to become a physician because that seemed to be the profession that would best position him for studies of sexuality. Between 1897 and 1928 he produced a seven-volume series on the topic, *Studies in the Psychology of Sex.* When first published, these books were condemned as being "filth," and after the first printing no more were permitted to be published in Great Britain. The subsequent ones, published in the United States, were available (legally) only to physicians. Like his iconoclastic friend George Bernard Shaw, Ellis was prone to advocate for ideas that were far ahead of his time. He proposed the creation of a national health service for Great Britain, preached the value of sex education, and was an ardent advocate for equal rights for women. Ellis was creatively gifted, and he was also aware that he had had emotional problems during his early years. Thus it was only natural that he would become interested in the relationship between creativity and mental illness, although he is much less well known for his contributions to that area of knowledge.

In *A Study of British Genius* (1926) Ellis attempted to provide an objective and unbiased examination of the associations between genius and psychopathology. Inspired by the work of Galton, but working empirically rather than to prove a point, Ellis provided the first objective quantitative study of the "genius and insanity" question. His method was to begin with the sixty-six volumes of the British *Dictionary of National Biography (DNB)*, roughly the equivalent of the American *Who's Who*. The edition of the *DNB* with which Ellis worked contained detailed summaries of the life and work of more than 30,000 British people. Well aware that most of these people were eminent persons rather than geniuses, he strove to arrive at a method by which he could whittle the list down to a reasonable number and also identify those who "displayed any very transcendent degree of native ability." He employed two basic principles to reduce the list to a final group of 1,030 people.

The first principle was that no person could be chosen simply for being a child of members of the nobility or the royal family, although he did include people who were from the "Commons" and who later achieved a peerage or who had children who were peers. His second major criterion was that the individual's biography occupied more than three pages in the *DNB*.

Having set these criteria, he went through the entire *DNB* and made some minor modifications to these two rules, especially the second. The primary modification was to include some people who had entries of less than three pages and who were nonetheless highly creative. Jane Austen was such an example. He also excluded people who were described for three pages or more who happened to play a significant role in history even though they had no extraordinary abilities, such as the regicide John Bradshaw, the lawyer who was president of the Parliamentary Commission that tried and sentenced King Charles I, thereby initiating Cromwell's short-lived Protectorate. Describing these rules and methods, Ellis said: "I have sought to subordinate my own private judgment in making the selection. It has been my object to place the list, so far as possible, on an objective basis." One might criticize his sample for many reasons (for example, the biographies were based on secondary information, limited to British figures, and more likely to have more accurate and complete information for those who lived more recently). But Ellis nonetheless also set new standards for objectivity in addressing the "genius and insanity" question.

Ellis's findings are summarized in Table 4.1. As is evident from the terminology of the table, which is drawn directly from Ellis's, he was working in a time when psychiatric diagnostic classifications and terminology were still relatively primitive and rudimentary. Further, he was working in a pre-epidemiological era, and so therefore he had no well-validated population norms against which he could compare the rates of observed psychopathology. The rates that he gleaned by poring over the *DNB* are likely to be underestimates. The meaning of some of Ellis's statistics is difficult to interpret, such as the high rate of imprisonment. This may reflect the fact that some eminent people became so because of involvement in politics (for example, Oliver Cromwell), and this may have predisposed them to imprisonment because of their unpopular views. Others (such as Oscar Wilde) may have been imprisoned because they exhibited sexual mores incompatible with the prevailing view of their era (in Wilde's case, homosexuality). But others may truly have had criminal personalities.

The rate of "insanity" noted by Ellis is certainly higher than is usually recorded for the general population, for which the current base rate is 1 percent for schizophrenia and 1 percent for mania. These are the two most common psychotic illnesses. The rate of melancholia—or what we currently call depression—is similar to current lifetime population rates of approximately 10 to 20 percent. (Rates vary depending on whether narrower or broader definitions are used.) Due to underre-

TABLE 4-1. Havelock Ellis: A Study of British Genius

1,030 Individuals cited in the *Dictionary of National Biography*

	N	%
"Insane"	44	4.3
Menlancholic	85	8.2
Shy, Nervous	68	6.6
Stutterers	13	1.2
Involuntary Tics	7	0.7

porting, however, Ellis's number may represent an underestimate. It is interesting that he also chose to examine the rates of other central nervous system symptoms such as stuttering and tics. It is even more interesting that in the preneuroscience era in which he was writing, he clearly recognized these phenomena as manifestations of pathology in the brain.

After the work of Ellis, the next major study that attempted to use objective methods was conducted in Innsbruck by the Austrian psychiatrist Adele Juda. Between 1927 and 1943, she conducted interviews and examined historical documents in order to evaluate the relationship between "genius and mental illness" in German-speaking artists and scientists of the eighteenth and nineteenth centuries. Like Ellis, she culled through large numbers in order to obtain an appropriate sample of creative people. She distilled the records of 19,000 people down to a group of 294 highly gifted individuals. She divided these into two categories: 113 artists and 181 scientists.

The results of her study are summarized in Table 4.2. The group of artists consisted of 12 architects, 18 sculptors, 20 painters, 26 musicians, and 37 poets. The most common mental illnesses that she observed in the artists were in a category that she referred to as "psychopaths." The diagnoses that she included in this group are somewhat similar to those which Ellis referred to as "nervous, shy, or emotional." Juda's psychopaths manifested traits such as being schizoid, eccentric, emotionally unstable, excitable and high-strung, hysterical, or weak of character. This category might be considered comparable to the present-day *Diagnostic and Statistical Manual (DSM)* category of personality disorder, or what used to be referred to as "neurosis." In this table, some diagnoses that would not be considered psychoses today, such as epilepsy, have been excluded, so that the numbers do not add up precisely to the total Ns of 113 and 163. The rates for psychotic illness were much lower: 2.7 percent schizophrenia, 2.7 percent underdetermined psychosis, and no manic-depressive illness. The patterns of psychopathology are somewhat different among the scientists. A higher number, 75.0 percent, were normal, as compared with 63.7 percent of the artists. Personality disorders were less common among the scientists: 15.3 percent. The rates for schizophrenia and manic-depressive illness were reversed in the scientists: 4.3 percent had manic-depressive illness, whereas none had schizophrenia. There was also a 2.4 percent rate of undetermined psychosis.

Juda subdivided the scientists into 51 from theoretical sciences and 112 from natural sciences. The vocational composition of these groups is somewhat strange by modern standards: "theoretical" sciences included history, jurisprudence, education, language, philosophy, and theology; "natural" sciences included astronomy, botany, chemistry, geology, mathematics, medicine, mineralogy, physics, and zoology. (She also included 18 statesmen and explorers in the total of 181. These are excluded from the table because they are so far removed from the modern concept of science.)

Juda's study is a landmark in the effort to identify relationships between creativity and mental illness, because of her large sample, her examination of both artists and scientists, and her meticulous efforts to catalogue specific kinds of disorders. Yet her findings are somewhat

TABLE 4-2. Adele Juda: The Relationship Between Highest Mental Capacity and Psychic Abnormalities

	N	%
Artists (N = 113)		
Normal	72	63.7
Manic Depressive	0	0
Schizophrenia	3	2.7
Undetermined Psychosis	3	2.7
Personality Disorders	31	27.4
Scientists (N = 163)		
Normal	124	76.0
Manic Depressive	7	4.3
Schizophrenia	0	0
Undetermined Psychosis	4	2.4
Personality Disorders	25	15.3

difficult to interpret. As in the case of Ellis, her terminology is antiquated, and it is not clear how her diagnoses translate to contemporary ones. She relied heavily on secondhand sources, as all of her subjects were drawn from the eighteenth and nineteenth centuries. And, finally, her group of scientists includes 51 in "theoretical" sciences that would very likely be classified as humanities disciplines today. The most striking oddity is in the contrasting rates of schizophrenia and manic-depressive illness in artists versus scientists. As we shall see below, nearly all subsequent studies have indicated a reverse pattern, or at least a higher rate of mood disorders in artists.

Improving Diagnostic Precision: The Iowa Writers' Workshop Study and the Quantitative Era

Although I am now a middle-aged neuroscientist, I was once a young English professor at the University of Iowa. Iowa's literature program was and is famous both nationally and internationally. Its Writers' Workshop is the jewel in its crown. The Iowa Writers' Workshop was founded in the 1940s by poet and former Rhodes Scholar Paul Engle. It was the first program in the United States to award a doctoral degree for creative writing. Its students and faculty are a veritable Who's Who of contemporary American literature, including (to name only a few) Kurt Vonnegut, Jori Graham, Gail Godwin, John Irving, Robert Lowell, Anthony Burgess, Mark Halperin, Mark Strand, Robert Coover, Philip Roth, and John Cheever. As a young faculty member in what was then a small and cohesive department, I mingled with the workshop crowd as well as the "lit crit" (literary criticism) crowd. My own interest in creativity and the brain probably even antedates that era, but my opportunity to unify those two domains of knowledge began to achieve realization after I decided to enter medical school after three years of teaching English at Iowa. I was drawn to medicine in part because I perceived it to have a greater social utility and in part because it would draw me into the vast and eternally dynamic world of science that contrasted sharply with the more static world of Renaissance literature and history.

Even before I entered medical school I had a feeling that brain science, or neuroscience as we have come to call it during the past twenty to thirty years, would be my first love, and that my medical specialty would focus on the study and treatment of brain diseases. So I decided to become a psychiatrist. The literary world from which I had come was well populated with people who had vividly described symptoms of mental illness. Here are just a few examples, from some of my favorite poems, representing many different periods of history.

There is Macbeth, deeply depressed after he begins to understand the consequences of his crime, and its punishment, in his troubled conscience:

> Life's but a walking shadow, a poor player,
> That struts and frets his hour upon the stage,
> And then is heard no more. It is a tale
> Told by an idiot, full of sound and fury,
> Signifying nothing.
>
> (William Shakespeare, *Macbeth*)

There is John Keats, the great Romantic poet who died at age 25, describing the nature of sadness and depression in his *Ode on Melancholy*:

> She dwells with Beauty—Beauty that must die;
> And Joy, whose hand is ever at his lips
> Bidding adieu; and aching Pleasure nigh,
> Turning to poison while the bee-mouth sips:
> Ay, in the very temple of Delight
> Veil'd Melancholy has her Sovran shrine,
> Though seen of none save him whose strenuous tongue
> Can burst Joy's grape against his palate fine;
> His soul shall taste the sadness of her might,
> And be among her cloudy trophies hung.
>
> (John Keats, *Ode on Melancholy*)

There is Gerard Manley Hopkins, describing his depressive self-hatred with intense poignancy at the conclusion of one of his sonnets:

I am gall, I am heartburn. God's most deep decree
Bitter would have me taste; my taste was me;
Bones built in me, flesh filled, blood brimmed the curse.
Selfyeast of spirit a dull dough sours. I see
The lost are like this, and their scourge to be
As I am mine, their sweating selves; but worse.

(Gerard Manley Hopkins, *I wake and feel the fell of dark*)

There is Emily Dickinson, who wrote an amazing body of poetry that was little known during her lifetime, but which astonishes us today with its innovation and richness of metaphor. Here is her description of pain—certainly psychological, and perhaps also physical:

Pain—has an Element of Blank—
It cannot recollect
When it began—or if there were
A time when it was not—

It has no Future—but itself—
Its Infinite contain
Its Past—enlightened to perceive
New Periods—of Pain

(Emily Dickinson, poem 650)

There is Robert Frost, a great American poet from New England, who is often thought to be simple and folksy, but who in fact had a very deep and dark side:

I have been one acquainted with the night,
I have walked out in rain—and back in rain.
I have outwalked the furthest city light.

I have looked down the saddest city lane.
I have passed by the watchman on his beat
And dropped my eyes, unwilling to explain.

I have stood still and stopped the sound of feet
When far away an interrupted cry
Came over houses from another street,

But not to call me back or say good-by;
And further still at an unearthly height,
One luminary clock against the sky

Proclaimed the time was neither wrong nor right.
I have been one acquainted with the night.

(Robert Frost, *Acquainted with the Night*)

By my senior year in medical school, I began to conceive of a study of the relationship between creativity and mental illness. I decided to conduct a careful and well-designed study of creativity and psychopathology using the faculty of the Writers' Workshop.

The work by Lombroso, Galton, Ellis, and Juda was a starting point, but the many deficiencies of those studies were obvious. What was needed was a study that used in-person, structured interviews and that applied modern psychiatric diagnoses using prespecified diagnostic criteria. This study was designed and initially executed in the early 1970s, well before the American Psychiatric Association (APA) developed its criterion-based *Diagnostic and Statistical Manual III (DSM-III)*. So I wrote a predefined set of criteria that were not dissimilar to the ones ultimately adopted by the APA.

In addition to incorporating diagnostic criteria, the Iowa Writers' Workshop Study also improved on its predecessors by including a group of educationally matched controls. The Writers' Workshop has a limited number of permanent faculty members (typically two poets and two prose writers). The remainder of the faculty in any given year consists of visiting writers who come to Iowa, drawn by its pastoral tranquility and an opportunity to be "far from the madding crowd" for a time of introspection, incubation, and isolation. Typically two or three such eminent writers are in residence in a given year, so this was the reservoir from which I drew my sample. Because the pool collected slowly over time, it took several years to execute the first published study, which consisted of fifteen workshop writers and fifteen educationally and age-matched comparison subjects who pursued occupations that did not require high levels of creativity.

I began the study with a perfectly reasonable working hypothesis. I anticipated that the writers would be, in general, psychologically healthy, but that they would have an increased rate of schizophrenia in their family members. This hunch made good sense, based on the information that I had at that time. I was influenced by my knowledge about people such as James Joyce, Bertrand Russell, and Albert Einstein, all of whom had family members with schizophrenia. Juda's data also suggested a possible association between artistic creativity and schizophrenia.

In addition, one of my colleagues at Iowa, Leonard Heston, had done a study of adopted children who grew up without knowing their biological mothers. Adopted children of mothers with schizophrenia were compared to adopted children of mentally normal mothers. "Adopted offspring studies" are an ideal experimental tool for separating the influence of genes from the influence of environment. Heston's work supported a strong genetic influence. The children of schizophrenic mothers had a 10 percent rate of schizophrenia, despite growing up in a normal environment. The adopted children of mentally healthy mothers had a rate less than 1 percent—the rate that is found in general population samples. While conducting the study, Heston made a serendipitous observation that the mentally healthy adopted children of schizophrenic mothers tended to have a higher rate of creativity than the adopted children of the mentally healthy mothers. He and I discussed from time to time the possibility that the genetic tendency toward schizophrenia could be expressed as creativity in a *form fruste* (that is, a mild but normal variant of the illness). Further, a psychiatrist in Iceland, J. L. Karlsson, also had looked into the genetic relationship between schizophrenia and creativity, by examining the relatives of individuals listed in Iceland's *Who's Who*. He reported that the relatives of these successful people had an increased rate of schizophrenia.

As I began to interview the writers and to administer a group of psychometric tests, the error of my working hypothesis rapidly became evident. To my surprise, the majority of the writers described significant histories of mood disorder that met diagnostic criteria for either bipolar illness or unipolar depression, and most of them had been treated for it. Some had required hospitalization, while others had re-

Table 4-3. The Iowa Writers' Workshop Study: Psychiatric Illness in 30 Writers versus 30 Controls*

	Writers		Controls			
	N	%	N	%	X^2	P
Bipolar I	4	13	0	0	–	ns
Bipolar II	9	30	3	10	2.60	ns
Unipolar	11	37	5	17	2.13	ns
Any Bipolar Disorder	13	43	3	10	6.90	0.01
Any Mood Disorder	24	80	9	30	13.20	0.001
Alcoholism	9	30	2	7	4.01	0.05
Drug Abuse	2	7	2	7	–	ns

*Some people had more than one diagnosis, so the numbers add up to more than 30. The last two columns are statistical tests of how significant the differences are. A P greater than .05 is considered statistically significant.

ceived outpatient medication or sometimes psychotherapy. Not a single writer displayed any symptoms of schizophrenia! When I submitted my results on the first fifteen writers and controls for publication, I had difficulty finding a journal that would accept the article. The reviewers selected to evaluate my submission apparently held the same preconceived notions that I had only recently rejected myself, and they did not welcome evidence to the contrary. Eventually, however, the paper saw the light of day, and I was subsequently able to expand the sample to a total of thirty writers and thirty controls over the next few years. The results of the later report are shown in Table 4-3.

Conducting that study was one of the most enjoyable experiences of my research career. Although many writers had had periods of significant depression, mania, or hypomania, they were consistently appealing, entertaining, and interesting people. They had led interesting lives, and they enjoyed telling me about them as much as I enjoyed hearing about them. Mood disorders tend to be episodic, characterized by relatively brief periods of low or high mood lasting weeks to months, interspersed with long periods of normal mood (known as euthymia to us psychiatrists). All the writers were euthymic at the time that I interviewed them, and so they could look back on their periods of depression or mania with considerable detachment. They were also able to describe how abnormalities in mood state affected their creativity. Consistently, they indicated that they were unable to be creative when either depressed or manic.

In a sense, the Workshop study confirmed two apparently conflicting, but prevailing, ideas about the nature of creativity and its relationship to mental illness. One point of view, for which the Terman study might be considered a prototype, is that gifted people are in fact supernormal or superior in many ways. My writers certainly were. They were charming, fun, articulate, and disciplined. They typically followed very similar schedules, getting up in the morning and allocating a large chunk of time to writing during the earlier part of the day. They would rarely let a day go by without writing. In general, they had a close relationship with friends and family. They manifested the Freudian definition of health: *lieben und arbeiten,* "to love and to work." On the other hand, they also manifested the alternative common point of view about the nature of genius: that it is "to madness near allied." Many definitely had experienced periods of significant mood disorder. Importantly, though handicapping creativity when they occurred, these periods of mood disorder were not permanent or long-lived. In some instances, they may even have provided powerful material upon which the writer could later draw, as a Wordsworthian "emotion recollected in tranquility."

The Writers' Workshop Study was followed by several others that also documented a relationship between mood disorders and creativity. One was done by psychologist Kay Jamison, who examined forty-seven poets, playwrights, novelists, biographers, and artists in Great Britain. Although she did not use a structured diagnostic interview or make diagnoses based on diagnostic criteria, she did obtain detailed information about history and type of treatment. These data, shown in Table 4.4, document that a substantial portion of these prominent British humanists had very high rates of mood disorder. More than 38 percent of the total sample had been treated for an affective illness, with the highest rate occurring in playwrights and the second highest rate occurring in poets. Thus the early Writers' Workshop data were solidly con-

TABLE 4-4. Kay Jamison's Study of 47 British Writers

	% Treated for Bipolar Illness (Hospitalization, lithium, ECT, etc.)	% Treated with Anti-depressants for Depression	% Treated with Psychotherapy Alone for Depression	Total % Treated for an Affective Illness
Poets	16.7	33.0	5.5	55.2
Playwrights	0.0	25.0	37.5	62.5
Novelists	0.0	25.0	0.0	25.0
Biographers	0.0	20.0	0.0	20.0
Artists	0.0	12.5	0.0	12.5
Total Sample	6.4	23.4	8.5	38.3

firmed in a second study collected by using direct interviews rather than relying on secondary sources. Although Jamison did not collect a control group for comparison, the rates are sufficiently similar to the Workshop study to be convincing.

Additional support for the association between mood disorders and creativity came from the work of Harvard psychiatrist Joseph Schildkraut, who studied fifteen abstract expressionist artists of the New York School who worked in the mid-twentieth century. Consistent with the Iowa Workshop study and the Jamison study, he found that about 50 percent of these artists had some form of psychopathology, which was predominantly mood disorder. Problems with alcohol abuse were also common, as also was the case in the Workshop study. This group of artists also suffered premature death, with nearly 50 percent dying before the age of sixty, two by suicide and two in single-vehicle accidents while driving. Although not as systematic as the Iowa Workshop or the Jamison study, Schildkraut's work provides additional evidence that there is a relationship between artistic creativity and mood disorders.

Is There a Connection between Creativity and Schizophrenia?

What about the association between creativity and schizophrenia, the hypothesis that launched my Writers' Workshop study? The answer to this question is uncertain at present. The evidence supporting an association between artistic creativity and mood disorder is quite solid, as is the absence of an association with schizophrenia. The nature of artistic creativity, particularly literary creativity, is probably not compatible with the presence of an illness like schizophrenia, which causes many of its victims to be socially withdrawn and cognitively disorganized. An activity such as writing a novel or a play requires sustained attention for long periods of time and an ability to hold a complex group of characters and a plot line "in the brain" for as long as one or two years while the novel or play is being designed, written, and rewritten. This type of sustained concentration is extremely difficult for people suffering from schizophrenia.

Creativity in other fields may, however, be compatible with an illness like schizophrenia, particularly those fields in which the creative moment is achieved by flashes of insight about complex relationships or by exploring hunches and intuitions that ordinary folk might find strange or even bizarre.

We are still in the anecdotal era with respect to studying the relationship between scientific creativity and mental illness, particularly schizophrenia. Anecdote does suggest that this could be fertile ground. For example, the family history of the famous twentieth-century mathematician and philosopher Bertrand Russell supports the hypothesis that scientific creativity may be linked to schizophrenia. Bertrand Russell had an interesting and inspiring life. He was linked to one of the great noble families of Britain, although this was a piece of his heritage that he never exploited. He lost both of his parents at the age of three and was reared by relatives. During his childhood he was an aloof, lonely, and somewhat insecure child. He attended Cambridge, where his precocious brilliance was immediately recognized and valued. He graduated as a "senior wrangler" (the student who achieves the highest score on the mathematics final exams), that pool of elite students on whom Galton based a portion of his study of hereditary genius. During his early twenties, Russell developed a new way of thinking about mathematics, known as "logical atomism." Working with his friend

Alfred North Whitehead, he produced a seminal work on mathematics, the three-volume *Principia Mathematica* (1903). It was immediately recognized as a masterpiece. His thinking bridged mathematics and philosophy, and he also contributed to the development of logical positivism. The influential philosopher Ludwig Wittgenstein, for example, was much influenced by Russell's thinking. Like many people in the sciences, Russell was an ardent political activist. He participated in multiple antiwar marches, for which he was fined and imprisoned on two occasions. (He may well have been one of Ellis's 1,030 "men of genius" selected from the British *DNB,* and he may well have been one of the 59 listed in the "imprisoned" category.)

Although he had a lonely childhood, an unstable personal life (three divorces in an era when divorce was uncommon), and a somewhat unusual adult personality, there is no indication that Russell himself suffered from schizophrenia. He illustrates the connection between mathematical genius and schizophrenia primarily by virtue of his family history. His family pedigree was loaded with people who suffered with definitively diagnosed schizophrenia or who received diagnoses that suggest schizophrenia. His uncle William was "insane." His aunt Agatha was delusional. His son John was diagnosed as having schizophrenia, and his granddaughter Helen suffered from schizophrenia and committed suicide by setting fire to herself.

Like Bertrand Russell, other eminent scientists have life histories that suggest that there may be a link between scientific creativity, particularly in mathematics and physics, and schizophrenia. Isaac Newton is another interesting anecdotal case. He was born prematurely, with tenuous survival during early infancy. Throughout his life he manifested a variety of schizotypal traits. He was chronically suspicious. He had a variety of unusual beliefs and interests that were extreme even for his time, such as an obsession with alchemy and the occult. He never married and lived alone most of his life. These traits alone would suggest that he falls within the "schizophrenia spectrum," but the diagnosis seems to be confirmed based on a psychotic break that developed at age forty. During this psychotic period, he suffered from delusions of being

John Nash

persecuted by others, as can be discerned in his correspondence with a variety of people, including Robert Hooke.

Many readers will be familiar with the history of John Nash, as described by Sylvia Nasar in *A Beautiful Mind* (1998) and portrayed in the Oscar-winning film of the same name. In addition to his contributions to game theory, for which he was awarded a Nobel Prize, Nash was also an original and gifted mathematician. Nash also exhibited schizotypal traits early in life and developed a psychosis at age thirty. His son, also named John, suffers from schizophrenia as well.

Albert Einstein had an unusual and eccentric personality and manifested many schizotypal traits, such as poor grooming and hygiene, and deficiencies in interpersonal skills. These traits might simply be written off as unimportant and a manifestation of his "genius," but for the fact

that his son by his first marriage (to Mileva Maric) suffered fr
ophrenia.

It is now time to build on this anecdotal information and to co-duct a study of gifted scientists, using the same type of design that I used for the Writers' Workshop study. The major obstacle to such work is difficulty in obtaining research funding, because the topic "falls between the cracks" for most funding agencies. Through a stroke of good fortune, however, a generous private donor has contributed a sum of sufficient size to permit me to begin this study, using updated technological advances such as neuroimaging. I suspect that this study will confirm the hypothesis that launched the Workshop study: the scientists themselves will not have serious mental illnesses, but they will have an increased rate of schizophrenia in their family members.

Mental Illness, Creativity, and the Brain

How might the brain forge a link between mental illness and creativity?

In chapter 2, I described many personality characteristics of creative people that make them more vulnerable, including openness to new experiences, a tolerance for ambiguity, and an approach to life and the world that is relatively free of preconceptions. This flexibility permits them to perceive things in a fresh and novel way, which is an important basis for creativity. But it also means that their inner world is complex, ambiguous, and filled with shades of gray rather than black and white. It is a world filled with many questions and few easy answers. While less creative people can quickly respond to situations based on what they have been told by people in authority—parents, teachers, pastors, rabbis, or priests—the creative person lives in a more fluid and nebulous world. He or she may have to confront criticism or rejection for being too questioning, or too unconventional. Such traits can lead to feelings of depression or social alienation. A highly original person may seem odd or strange to others. Too much openness means living on the edge. Sometimes the person may drop over the edge . . . into depression, mania, or perhaps schizophrenia.

We have seen how creative ideas arise in the brain—that violent and energetic process so well described in *Kubla Khan:*

> And from this chasm, with ceaseless turmoil seething,
> As if this earth in fast thick pants were breathing,
> A mighty fountain momently was forced:
> Amid whose swift half-intermitted burst
> Huge fragments vaulted like rebounding hail,
> Or chaffy grain beneath the thresher's flail:
> And 'mid these dancing rocks at once and ever
> It flung up momently the sacred river.

And we have seen how creative ideas probably occur as part of a potentially dangerous mental process, when associations in the brain are flying freely during unconscious mental states—how thoughts must become momentarily disorganized prior to organizing. Such a process is very similar to that which occurs during psychotic states of mania, depression, or schizophrenia. In fact, the great Swiss psychiatrist Eugen Bleuler, who gave schizophrenia its name, described a "loosening of associations" as its most characteristic feature: "Of the thousands of associative threads that guide our thinking, this disease seems to interrupt, quite haphazardly, sometimes single threads, sometimes a whole group, and sometimes whole segments of them."

When the associations flying through the brain self-organize to form a new idea, the result is creativity. But if they either fail to self-organize, or if they self-organize to create an erroneous idea, the result is psychosis. Sometimes both occur in the same person, and the result is a creative person who is also psychotic, as John Nash was for many years. As Nash once said: "the ideas I have about supernatural beings came to me the same way that my mathematical ideas did, so I took them seriously."

✓ Delusions—fixed false beliefs—are a very common symptom of psychosis. Typically delusions involve misinterpretations or misperceptions about things going on around a person. For example, a person may believe that a neighbor is transmitting messages into his brain or that a family member is poisoning his food. Somehow, in a delusionally psychotic person, the associations around the neighbor have become misconnected, so that a neutral or benign person is perceived to be malevolent. Or the associations with the family member and food become erroneously confused and misunderstood. Sometimes delusions begin vaguely and then become quite specific and fixed. There is a term for this in psychiatry: the delusions are said to "crystallize." The process may be much like those flashes of insight that lead to a creative outcome, but in this instance the result is instead a serious symptom of mental illness.

Another brain mechanism that may be common to mental illness and creativity is a problem with filtering or gating the many stimuli that flow into the brain. This is sometimes referred to generically as an *input dysfunction,* or a problem with filtering or sensory gating. All human beings (and their brains) have to cope with the fact that their five senses gather more information than even the magnificent human brain is able to process. To put this another way: we need to be able to ignore a lot of what is happening around us—the smell of pizza baking, the sound of the cat meowing, or the sight of birds flying outside the window—if we are going to focus our attention and concentrate on what we are doing (in your case, for example, reading this book). Our ability to filter out unnecessary stimuli and focus our attention is mediated by brain mechanisms in regions known as the thalamus and the reticular activating system.

Creative individuals (particularly the writers in my Workshop study) have sometimes complained that they are too easily flooded by stimuli, so that they become easily distracted. Some writers found that they were prone to be too sociable, such that their tendency to seek out other people interfered with getting their work done. They often had to organize their lives so that they were isolated from human contact for long blocks of time. In fact, the tendency of some to drink excessively may have been an effort to use alcohol as a central nervous system depressant to cope with their sensitivity to being flooded by stimuli. Being overwhelmed by more stimuli than the brain can manage is a potential mechanism that could lead to manic highs. During a manic episode people become excessively energetic, distractible, talkative, and full of ideas. A manic high is usually followed by a depressive crash. Depression is an alternative mechanism for dealing with an input dysfunction. In this case the individual copes by withdrawing from social contacts and almost everything else, even food and sex.

Sensitivity to excessive inputs from the fives senses may also, however, be a resource for creativity. Someone who takes in more stimuli than the average person is likely to experience more of life, to have greater awareness of feelings and needs, and to have more unusual perceptions and feelings than someone who can easily filter out stimuli that are inconvenient or noisy. Unlike schizophrenia, mania and depression are episodic illnesses. People who suffer from them usually recover fully and have a normal mood most of the time. During these prolonged euthymic periods the creative person may be able to organize the wealth of experiences and insights from episodic imbalances and to use them as a resource on which to draw when creating a poem or a painting. Because creativity in the arts is more closely linked to mood disorders, having difficulty in modulating the floods of stimuli may be a more important brain mechanism for this type of mental illness.

What Are the Effects of Treating Mental Illness in Creative People?

Some creative people have concerns about receiving psychiatric treatment, fearing that their creativity might be impaired. Some fear that psychotherapy, particularly psychoanalytic therapy, might "mess with their minds," perhaps by normalizing them and removing the neurotic habits or experiences that might be driving them to be creative. Another fear is that treatment with medications might take away or compromise the wellsprings of creativity, since medications so obviously have direct effects on the brain.

One can also hypothesize that mental illness, at least in the form of mood disorder, may potentially confer some small benefits. The increased energy that occurs during mania or hypomania could potentially enhance creativity. Further, the tendency to have enriched associative thinking (whether "on target" or bizarre) is also a component of mental illness. Ideas generated during hypomanic or psychotic states can potentially be explored and evaluated again when the mood is more neutral or the mind more rational. Some people have romanticized and idealized mental illness, seeing it as a reservoir of intense inner experience from which the creative individual can draw.

Yet there are often compelling reasons why treatment is necessary, and why many creative people welcome having their symptoms relieved or diminished. Both mood disorders and schizophrenia have unusually high rates of suicide. Many creative individuals have taken their lives. Notable examples include Vincent van Gogh, Virginia Woolf, Ernest Hemingway, Sylvia Plath, and Ann Sexton. In most cases these deaths could have been prevented with adequate treatment. Vincent, Virginia, and the others could have continued to enrich our world through their paintings and writings.

In the twenty-first century we are in a much better position to treat mental illnesses than in the time of Vincent or Virginia. We are fortunate that a large number of effective medications have been developed. The current situation is in stark contrast to the plight of the mentally ill only fifty years ago, when no treatments were available and when the majority of people were condemned to suffer through episodes of mania or depression until they spontaneously remitted. Those with schizophrenia tended to remain severely and chronically ill. In the early 1950s new medications were discovered. These included tricyclic antidepressants such as imipramine and neuroleptic medications for schizophrenia such as thorazine. Since that time steady improvements have been made in treatments both for mood disorders and for schizophrenia. The efficacy of lithium carbonate for the treatment of mania was established in the 1970s. Thereafter other mood stabilizers were introduced, such as depakote, for the treatment of mania. Newer antidepressants and neuroleptics have also been developed during the past decade or two; in general, these have fewer side effects that might affect creativity, such as sedation.

However, we have very few good scientific studies of their effects on creativity. The single best study thus far was published in 1979 by

Mogens Schou, a pioneer in the development of lithium carbonate as a treatment for mania. He studied a group of artists who suffered from bipolar disorder and measured their productivity and the quality of their work. He classified them into three groups based on their response to lithium. One group showed a great improvement in productivity. These were people who had very severe bipolar disorder and who found that lithium improved their ability to be creative. The second group experienced little or no effect on their creativity: they were neither better nor worse. Schou suspected that they probably were not taking their medication. The third group showed a decline in creativity. These people tended to rely on insights gained during a "manic high" to enhance their creativity. For this third group, the crucial issue is whether there is a tradeoff between the benefits that occur when manic symptoms are controlled and the reduction in creativity that occurs as a consequence of treatment. This is an issue that is likely to be settled on a case-by-case basis.

The case of Robert Lowell, one of America's finest twentieth-century poets, provides an interesting illustration of the effects of lithium on creativity. Lowell suffered from severe bipolar disorder that required multiple hospitalizations. His symptoms were only minimally helped by antidepressants or neuroleptics. He received extensive psychotherapy, which was also not helpful. When lithium carbonate became available in the late 1960s, he began to take it. He noticed a remarkable diminution in his mood swings that appeared to enhance his creativity. After beginning treatment in the spring of 1967, he was able to escape his annual breakdown and to write a prolific series of sonnet-like fourteen-line poems—seventy-four of them between June and December. The novelist Richard Stern describes Lowell's reaction in a journal entry dated 27 December 1968:

> He showed me the bottle of lithium capsules. Another gift from Copenhagen. Had I heard what his trouble was? "Salt deficiency." This had been the first year in eighteen he hadn't had an attack. There had been fourteen or fifteen of them over the past eighteen years. Frightful humiliation and waste. He'd been all set up to taxi up to Riverdale five times a week at $50/session, plus (of

course) taxi fare. Now it was a capsule a day and once/week therapy. His face seemed smoother, the weight of distress-attacks and anticipation both gone.

Lowell's literary biographer, Ian Hamilton, comments that the poems produced after lithium therapy show a "low key agreeableness." Thus the improved productivity and stability may have come at some expense in poetic power, although Lowell himself appears to have been most impressed with the extent to which lithium was helpful to him and apparently felt an improvement rather than a decline in his ability as a poet. His experience is consonant with observations of other creative individuals who have suffered from depression. For example, the great American composer Aaron Copeland has said, "Too much depression will not result in a work of art, because a work of art is an affirmative gesture." The painter Raphael Soyer also expresses this point of view: "I know that when I am depressed, it is harder for me to work." My Workshop writers also confirmed that they were usually unable to be creative during their periods of mood disorder.

Apart from Schou's study of lithium, however, there are no studies of the effects of medication on the creativity of artists or scientists. Such studies are difficult to do, because they require access to a group of creative people who suffer from mental illness. We thus remain limited to subjective accounts. In my own experience, most creative people feel that mental illnesses such as mania or depression are the enemy of the creative process. If their symptoms are severe or crippling, creative people want to have them treated. However, individual clinicians working with individual patients must proceed with care, sensitivity, and respect. They must walk a fine line between overtreating (too much medication, causing impairment in creativity) and undertreating (too little medication, resulting in the patient's being too symptomatic to be creative). The aim is to reduce symptoms and suffering without sacrificing creativity. The aim is also to ensure that a valuable life is not lost through suicide. The physician's motto—*primum non nocere,* "first, do no harm"—sets forth an important and challenging standard for those who are privileged to provide psychiatric care for creative people who suffer from mental illness.

5

WHAT CREATES THE CREATIVE BRAIN?

A devil, a born devil, on whose nature
Nurture can never stick! On whom my pains,
Humanely taken, all, all lost, quite lost!

—*William Shakespeare*

Francis Galton may have popularized the contrast between the influences of nature and nurture when he published a book with that title in 1874: *English Men of Science: Their Nature and Nurture.* But the original usage harks back to the lines above from Shakespeare's *The Tempest.* Prospero, a great "white magician" who is the conceptual ancestor of more modern wizards such as Gandalf in *Lord of the Rings* and Dumbledore in *Harry Potter,* is lamenting his hopelessly failed experiment with Caliban. Caliban is a brutish creature whom Prospero discovered when shipwrecked on the mythical island that forms the site of the drama. Prospero attempted to nurture and civilize him by adopting him as almost a son and letting him live as a member of the family. In return

Caliban attempted to rape Prospero's daughter Miranda, in the hope of populating the island with more Calibans. As Caliban says:

> You taught me language, and my profit on't
> Is, I know how to curse.
>
> (I.2.362–63)

The character of Caliban is contrasted with that of Ariel, a spirit of light who wants only to do good. The implicit message of this contrast seems to be that nature—or what we would now call heredity or genetic factors—is a powerful force that shapes what we become. In some instances, it is so powerful that no environmental factors can affect the direction in which genetic influences drive the individual.

The relative influences of genes and environment, or nature versus nurture, are still a lively topic of debate in the twenty-first century. Although some extremists might take a strong position at one or the other end of this polarity, insisting that genetic influences alone, or environmental influences alone, determine what we become, most thoughtful scientists are aware that these two forces work in dynamic interaction. That is, we are shaped by both.

The intriguing questions are: How? Why? When?

These questions are especially intriguing when we consider how the creative person emerges, how the creative process begins to occur, and how this is implemented as the creation of the creative brain.

The Role of Nurture: Cradles of Creativity

If we were to plot major creative contributions over human history, we would find that they tend to occur in clusters. Extraordinarily creative people usually do not emerge in a random way over time. During some periods in history, so far as we know, very few extraordinarily creative works or ideas were produced. During other periods, an exuberant flowering of the human creative spirit occurred. Individuals emerged from "nowhere" and created some of the highest achievements of the human mind and brain.

These time periods give us an historical laboratory in which we can investigate the role of the environment in nurturing creativity. We can learn a great deal by looking back at them and examining how creative brains might have been created.

Many eras could be chosen. Here are a few examples of these "cradles of creativity." There is fifth- and fourth-century (B.C.) Athens, a Greek city-state that established the first (intermittently) successful democracy. Early in the fifth century the Persians, led by Darius, had attempted to invade Athens and its allies within the Delian League. The Persians (whose country is now known as Iran) were defeated, and the Athenians began to rebuild the temples damaged in the war. The great leader Perikles began a major building program to beautify Athens. From this emerged perhaps the most perfect building ever created, the Parthenon, designed by the architects Iktinos and Kallikrates and decorated with friezes and statues made by the great sculptor Phidias. Completed in the 440s, it stood virtually intact until A.D. 1687, when the Venetian General Francesco Morosini bombarded it with cannon fire during a siege and exploded gunpowder that was stored inside. The roof was knocked off, and many of the friezes and sculptures fell to the ground, where they remained for many years. Some were stolen by anonymous thieves. Others were removed by Lord Elgin of Britain in the early nineteenth century and sold to the British government for £35,000. They were placed on display in the British Museum, where they are known as the "Elgin Marbles." Most of the Parthenon still stands high and proud on the Acropolis ("City on High") in Athens, with tasteful restoration by the Greek government. It is one of those "must see" sights for whomever can manage it. How did they ever chisel those huge columns and raise them to the sky with the relatively primitive tools available to them? How did they develop the mathematical proportions that give the building such a perfect appearance of symmetry? (Every aspect was built with a 9:4 ratio to produce perfect proportions.)

Athens was not just the cradle of democracy, the creator of magnificent architectural and engineering feats, and the source of humanistic and realistic depiction of the human face and body in sculpture as ren-

The Parthenon

dered by Phidias and his followers. It also fostered the creation of intellectual traditions that we know today as "Western philosophy." Athens was the home of Socrates, Plato, and Aristotle, a succession of teachers surrounded by students who helped advance their ideas in an "intellectual academy." Socrates (d. 399 B.C.) left behind no writings. The "writings" of Aristotle (d. 322) appear to be mostly very condensed notes taken by his students. Only Plato (d. 347) survives as the great literary and philosophical master that he was, creating lucid dialogues about the nature of reality, of the ideal government, of beauty and love, and many other topics. Their ideas influenced the development of Christian philosophy through followers such as Augustine or Thomas Aquinas, and they are echoed over and over in later philosophical writings by people such as Kant and Hegel. It has been said that the rest of Western philosophy is simply a series of footnotes to their ideas.

Athenians of the fifth century also produced great literature. Theater and drama as we know them were invented in Athens during this time. Again, much is lost, but works survive by the three great writers of tragedy—Aeschylus (d. 456), Sophocles (d. 406), and Euripides (d. 406)—as well as the great comedian Aristophanes (d. 386). These works explore timeless themes such as conflicts of responsibility to one's self and state or country, the relationship between men and women, and the rationale for fighting wars.

Another great cradle of creativity was Paris in the second half of the nineteenth century. In the domain of architecture, Baron Georges Haussmann, appointed by Napoleon III to modernize Paris, remade many parts of the medieval city. Beginning in the 1850s, he filled it with wide boulevards and open squares, creating a grand setting for its older architectural gems such as the Louvre, Notre Dame, and the Conciergérie. Many people consider Paris to be the most beautiful city in the world, and it is truly the "City of Light." In parallel with the flourishing of architecture, French painters of that era were revolutionizing the visual arts. Departing from the highly realistic neoclassical style, they began to portray people and objects based on the emotional impressions that were created in their minds, launching the artistic movement that we call impressionism. From this movement, banned from the Academy and relegated to the Salon de Refusés (Gallery of the Outsiders), emanated further experiments with portraying how the mind's eye sees: postimpressionism, pointillism, fauvism, expressionism, and cubism. Economic prosperity created by the Industrial Revolution permitted the building of great train stations in Paris, linking the city to the countryside and permitting artists to paint *en plein aire,* where they could study the varying effects of lighting. The painters of this era are probably the most beloved and approachable artists to most of us who live in the twenty-first century: Edouard Manet, Henri Toulouse-Lautrec, Claude Monet, Auguste Renoir, Paul Cézanne, Vincent van Gogh, Pablo Picasso, and many others.

Paris of the mid- and late nineteenth century was also a major center for opera, housed in the grand Opéra building designed by Charles Garnier during Haussmann's architectural reform of the city. Classics such as Georges Bizet's *Carmen,* Charles Gounod's *Faust,* and Antonio Rossini's *William Tell* premiered in Paris. Jacques Offenbach's *Orpheus in the Underworld* added a comic touch and introduced the can-can to the staid world of opera. In the domain of literature the French nineteenth-century writers conveyed a preoccupation with social justice through works such as Victor Hugo's *Les Misérables* and Emile Zola's *J'Accuse.*

The United States during the late nineteenth and early twentieth centuries was also a cradle of creativity, although in this case the creativity was more practical and less artistic. With their penchant for efficiency, Americans have given the world some of its most useful inventions. As school children, we all learned about how Eli Whitney's invention of the cotton gin transformed the economy of the South. Samuel Morse invented the telegraph and Morse code, creating one of the first mechanisms for rapid communication throughout the country and ultimately throughout the world. Alexander Graham Bell, although born in Scotland, lived most of his life in the United States and gave us another resource for rapid and efficient communication: the telephone. Thomas Edison created the light bulb, the phonograph, the storage battery, and "talking movies," as well as numerous less significant inventions. Although Henry Ford did not invent the automobile, he created the assembly lines that permitted them to be mass-produced and that allowed ordinary people to own a car. Although the Wright brothers perhaps did not create the first airplane, their success in getting a gas-propelled heavier-than-air plane literally off the ground in Kitty Hawk, North Carolina, in December 1903 paved the way for the modern world of air travel. The Henry/Winchester repeating rifle changed the world by making warfare more efficient, as did the Colt revolver. (In addition to appreciating American inventiveness, we should perhaps also mourn the fact that Americans have been so successful in inventing "weapons of mass destruction.")

On the intellectual side, William James created the philosophy of Pragmatism, providing a name for the practical-mindedness that shapes much of American thinking. He also could be considered one of the founding fathers of the discipline that we now call psychology, writing the classic *Principles of Psychology* (1890)—still one of the world's best reads on how the human mind is organized. His follower John Dewey created a philosophical framework for education and curricular development, emphasizing the importance of interaction with others, reflective thinking, preparation for living in a democracy, and vocational training. Sometimes referred to pejoratively as "robber barons" or more kindly as "captains of industry," individuals such as Andrew Carnegie and John D. Rockefeller created companies such as U.S. Steel, the Union-Pacific Railroad, and Standard Oil—demonstrating both the power of capitalism and also its spinoff in philanthropy. Their companies and others built the powerful economic infrastructure that permitted the United States to come to the aid of Europe during two World Wars. They also founded some of our major charitable foundations that work for the common good.

Other cradles of creativity have also been important in human history. There is fin-de-siècle Vienna, with Sigmund Freud, artistic secessionists such as Gustav Klimt and Egon Schiele, and the composer Gustav Mahler. The Enlightenment in France had Voltaire, Jean-Jacques Rousseau, and Denis Diderot. Tudor and Elizabethan England had Thomas More, Christopher Marlowe, Edmund Spenser, William Shakespeare, and John Donne. Late eighteenth-century America had Thomas Paine, Benjamin Franklin, and Thomas Jefferson. All these "cradles" have many features in common.

Renaissance Florence as a Laboratory for the Case Study of Nature and Nurture

We could explore any or all of these times and places as case studies in how environmental factors may influence the emergence of creative ideas in the human brain. But it is perhaps best to closely examine one that can be considered the most prototypical as a cradle of creativity, in order to identify how it nurtured great creative genius. Since I'm the author of this book, I'll take the liberty of choosing my favorite among these various cradles: the Renaissance, and especially its most important home, Florence. Florence was the city that launched two of the most fascinating and diverse geniuses of human history, Leonardo da Vinci and Michelangelo Buonarroti. How did this happen? Many of the great Renaissance artists came from very unlikely origins. Nothing was present in either their heredity or their early childhood environment that would predispose them to become creative or artistic or inventive. Yet someone identified them as having a natural gift and placed them in an environment that could nurture them.

Let us begin this effort to understand the interaction between nature and nurture in our Florentine laboratory by going back to the origins of the Italian Renaissance. Giorgio Vasari, himself an artist, was the great biographer of Renaissance artists, summarizing their lives in a book called *Lives of the Artists,* first published in 1550 and revised in 1568. In fact, Vasari coined the word "renaissance," or *il rinascimento,* which literally means "rebirth." His book is a valuable resource for learning about how the creative brain is created.

In dramatic language, Vasari dates the beginning of the Renaissance to the work of Cimabue:

> When the barbarian hordes devastated unhappy Italy, he says, they not only destroyed the buildings, but exterminated the artists themselves. Then, by the grace of God, in the year 1240, Giovanni Cimabue was born in the city of Florence to give the first light to the art of painting.

Like many Renaissance artists, Cimabue came from humble origins and began to study with local artists as a boy. He first imitated the Byzantine style that prevailed but gradually began to create more realistic paintings. His student Giotto (1266–1337) was perhaps the first true Renaissance artist, because he was the first to render naturalistic facial expressions and body postures, which he did in a breathtakingly beautiful manner. Vasari's account of Giotto's early origins contains themes that will be echoed over and over as we examine the various ways in which Florence became a cradle of creativity.

> The birth of this great man took place at Vespijnano, fourteen miles from Florence. His father, Bondone, was a simple farmer and brought up the clever, lively, likeable boy as well as he could. When Giotto was about ten years old,

Bondone gave him his sheep to watch, and with them the lad wandered, now here, now there, near the village. But, impelled by Nature herself, he was always drawing. On the stones, the earth, or the sand, he drew pictures of things he saw or of his fancies. It chanced, one day, that Cimabue happened to see the boy drawing the picture of one of his sheep on a flat rock with a sharp piece of stone. Halting in astonishment, Cimabue asked Giotto if he would go with him to his house. The child said he would go gladly if his father would allow it. Bondone readily granted his permission and his son went off to Florence. Under Cimabue's guidance and aided by his natural abilities, Giotto learned to draw accurately from life and thus put an end to the rude Greek (Byzantine) manner. He introduced the custom of drawing portraits, which had not been done for more than 200 years. Among his portraits were those of Dante Alighieri, his intimate friend.

Although not the multitalented polymath that Leonardo and Michelangelo were to become, Giotto was nonetheless a diversely gifted man. He is best known for his paintings done in fresco. (Fresco is a challenging technique where the paint is applied to a wet plaster surface and then dries intact. It allows no margin of error. It was a standard technique of Renaissance painting.) Some of these are comprised of multiple images that tell life stories. Among the greatest treasures are frescos showing the life of Saint Francis in Assisi and those showing the life of Jesus in the Arena Chapel in Padua. What was particularly new in Giotto's art was an extraordinary skill in portraying facial expressions and subtle emotions. For example, a small panel in the series in the Arena Chapel shows Judas kissing Christ after having betrayed him. Both faces tell a story. Jesus knows what is to come, and Judas is visibly duplicitous. Human emotion had not been rendered so realistically in art since classical times.

Giotto was not only a painter, but also an architect and sculptor. He designed the sublime campanile (bell tower) that stands beside the great cathedral of Florence, Santa Maria del Fiore, or the Duomo. There is a story about how Giotto came to be chosen to do the design. (Another version of this story is how he was chosen to paint frescos to decorate Saint Peter's Basilica in Rome for Pope Boniface VIII.) In either version there is a competition in which multiple artists participated. All others submitted complex designs. Giotto, however, simply drew freehand a circle so perfect and exact that it was considered a marvel to behold. With this simple and confident effort, he won the competition hands down. Such a bold sense of the strength of his art is typical of the great Renaissance artist. Long after Giotto's death Lorenzo the Magnificent, the greatest of the Florentine Medicis, decreed that a monument in Giotto's memory be placed in the Duomo with the inscription *I am he by whom the extinct art of painting was revived.*

Leonardo da Vinci and Michelangelo Buonarroti

The two greatest figures in Renaissance art—and perhaps the art of any era—are of course Leonardo da Vinci and Michelangelo Buonarroti. These two men, who were intense rivals, had a great deal in common. Both came from relatively simple origins, commercial backgrounds where there was no emphasis on creativity. Both, however, were driven from within by an internal predisposition ("nature") to express their creativity, initially as artists and later as scientists. Both were, like Cimabue and Giotto before them, given their first artistic mentoring as apprentices to great Florentine artists. Both were hugely gifted in many different areas. Both were careful students of nature. Both saw no boundaries between art and science. Both were highly inventive. Both worked intensely and passionately and for long hours at a time. Both were perfectionists and had problems with completing works with which they were dissatisfied. Both were financially supported in their efforts by wealthy noble families. Both lived long lives and were working on something right up to the end. And neither ever married or procreated. Their achievements in art and science were their legacy to the world.

It is impossible in this short book to describe all their many accomplishments. But we can examine enough to get some insight as to how their creative brains were created.

Leonardo (1452–1519) was the older of the two, older than Michelangelo by slightly more than one generation. He was the illegitimate son of Ser Piero da Vinci, a notary who enjoyed a modest social

Leonardo da Vinci

da Vinci's Angel, from *Baptism of Christ*

status in Vinci, a small town close to Florence. Leonardo had very little formal education. For example, he was taught no Latin or Greek, the languages of cultured men during this era. His mother was a peasant from nearby Anchiano, and Leonardo was raised by her family for his first five years. When Ser Piero's legal wife was found to be infertile, Leonardo was briefly moved into the patriarchal home, where his gift for drawing was noticed. He also was a talented musician, composing both words and music on the lute. Ser Piero took some of his son's drawings to Andrea del Verrocchio, one of the foremost Florentine sculptors and painters, to see if the work showed any promise. Verrocchio was astounded by the boy's ability and told Ser Piero that Leonardo should become a painter. Accordingly, Leonardo went to work in Verrocchio's studio.

At this time in Florence each of the great master artists ran a studio in which young men were apprenticed and trained. They would learn the various technologies of art, such as grinding colored stone to create paints of various colors, making brushes, shaping terra cotta models of statues, and applying the wet plaster used to create frescos. Leonardo quickly emerged as the most talented among the boys and young men who assisted Verrocchio. In fact, Vasari tells the following story:

> While he was in Andrea Verrocchio's shop, that master was engaged on a picture of Saint John Baptizing Jesus Christ. Leonardo painted an angel holding some vestments, and, although he was but a youth, the angel was the best part of the picture. This caused Verrocchio never to touch color again, so much was he chagrined to be outdone by a mere child.

Today that painting can still be seen. It hangs in the Uffizi Gallery in Florence. Leonardo's angel, the star of the show, stands out vividly at the left side of the picture. The delicate shading and rendering of space, *sfumato,* for which Leonardo's paintings are famous, is already apparent.

Unfortunately, all too little of Leonardo's prolific artistic output survives today. There are several reasons for this. One is that he was a perfectionist and often left works unfinished. Another is that he was prone to experiment with new techniques for applying paint colors, and some of these were not very successful, causing the paintings to deteriorate or even disappear. In one instance—a competition between Leonardo and his great rival Michelangelo—both created monumental renderings of battle scenes to decorate the Hall of the Grand Council in the Palazzo Vecchio (Old Palace) in Florence. Leonardo chose to try to use oil for his painting, and it was absorbed so quickly that the application failed. He simply abandoned the project, and nothing survives except some of his preliminary sketches.

Both his experimentation and his perfectionism also plagued the creation of *The Last Supper*—the painting for which, together with the *Mona Lisa,* he is most famous. This work was being done while Leonardo was in Milan, working for its Duke, Ludovico Sforza. Like many other Last Suppers in that era, it was located in the dining room of the monastery. Leonardo labored over rendering all of the faces of the disciples at the moment when they learn that one of them will betray Jesus. Each expresses complex and powerful emotions, such as bewilderment or grief. The two most important faces were those of Jesus and Judas. While he was working on this painting, Leonardo would often pause and look at the painting and simply think, sometimes for as long as a half day. According to Vasari, the Prior of the monastery, accustomed to ordering his nonmonastic employees to work for hours in his garden, complained to the Duke about Leonardo's apparent laziness and slowness. So Leonardo explained to the Duke what he was actually doing:

> He made it clear that men of genius are sometimes producing most when they seem least to labor, for their minds are then occupied in the shaping of those conceptions to which they afterward give form. He told the duke that two heads were yet to be done: that of the Saviour, the likeness of which he could not hope to find on earth and had not yet been able to create in his imagination in perfection of celestial grace; and the other, of Judas. He said he wanted to find features fit to render the appearance of a man so depraved as to betray his benefactor, his Lord, and the Creator of the world. He said he would still search but as a last resort he could always use the head of that troublesome and impertinent prior. This made the duke laugh with all his heart. The prior was utterly confounded and went away to speed the digging in his garden. Leonardo was left in peace.

Leonardo was in fact never able to finish completely the face of Jesus, because he could not figure out how to "render the divinity of the Redeemer." The anecdote illustrates two common traits of the creative mind and brain: the need to have free-floating periods of thought during which inspiration may come as the brain spontaneously self-organizes and new associative links are found, and the uncompromising and obsessional perfectionism that seeks to achieve the ideal product or result.

Leonardo excelled in many fields in addition to painting, sculpture, and music. He was also an engineer, an inventor, and a scientist. From the moment he moved to Florence from the rural countryside, he was devising schemes to re-engineer the Arno River or elevate the Church of San Giovanni by creating a flight of steps on which it would stand. His drawings show how he designed levers, cranes, dredging machines to clean riverbeds, and equipment to build tunnels through mountains. He famously drew a design (perhaps the first) for a human flying machine and attempted to build it. He presented himself to the Duke of Milan as a skilled designer of military equipment, and thereby won his patronage. He designed fortifications for cities and drew the first picture of a helicopter.

He was one of the first serious students of human anatomy. His anatomical drawings, which survive today and are in the possession of the English royal family, indicate that he spent many hours dissecting the human body. He was also fascinated with the anatomy of horses, animals that he loved and that figure prominently in his art. He thought three-dimensionally, as most in his era did not. One of his drawings illustrates the principal of tomography—of cutting through structures such as an arm or leg at regular spatial intervals, in order to create in one's mind the three-dimensional relationships of the components. This principle forms the basis for most of our modern neuroimaging tools, such as the earliest, computerized axial tomography, the "CAT scan." He was interested in mathematical proportions as well, as illustrated in his famous Vetruvian man.

In short, Leonardo is a model of the "Renaissance man." He excelled creatively in many different areas: art, music, engineering, and science. In our structured and compartmentalized modern world, we almost automatically assume that there is a dichotomy between art and science. A person must be in one domain or the other. But Leonardo illustrates that interest and skill in both can exist in perfect harmony and in fact that both enrich one another. Because he knew anatomy, he could mentally build the body of a man or a horse from the inside out. What nurtured such creative genius and such diversity? We shall see, shortly.

However, we must first go on to examine that second creative genius who was nurtured in the Florentine environment, Michelangelo Buonarroti (1475–1564).

Michelangelo's parents were solid citizens, but had no known interest in the arts. His father was mayor of a small town for a brief time and then settled in Settignano, near Florence. The major industry of Settignano was quarrying stone. Because his mother was in ill health, Michelangelo was given to a stonecutter's wife to nurse. Years later, he remarked to Vasari one day: "Giorgio, if I'm good for anything, it is because I was born in the good mountain air of your Arezzo and suckled among the chisels and hammers of the stonecutters." The Buonarroti family was not wealthy, and so Michelangelo was sent to school only briefly. He seemed more interested in drawing than studying. At thirteen or fourteen he was apprenticed in the shop of another of the major Florentine artists, Ghirlandaio. His gifts were immediately evident.

Michelangelo was particularly fortunate to be living during the time when Lorenzo de' Medici, known as the Magnificent, was the dominant figure in Florentine life. Lorenzo had acquired an impressive collection of classical sculpture, which was slowly being rediscovered and preserved during the Renaissance. His ambition, however, was to help create a "modern" school of sculptors who would recover the lost art of such sculpture and equal the achievements that were already occurring in Italian Renaissance painting. He asked Ghirlandaio to send some young men of promise to work in his sculpture garden. Michelangelo was one of them. Michelangelo first worked in terra cotta and then made a faun from marble. Lorenzo was so impressed by his skill that he invited Michelangelo to come to live in his princely household.

The child who nursed from a stonecutter's wife had stepped onto a new stage. Lorenzo's household was indeed magnificent. The best minds in Florence, known as the Florentine Academy, met there regularly to discuss their philosophical ideas. Michelangelo also met

Michelangelo Buonarotti

younger members of the Medici family, two of whom would later become Popes Leo X and Clement VII and commission him to beautify Saint Peter's Basilica in Rome. Young Michelangelo was surrounded by art from which he could learn and supplied with all the materials that he needed to sculpt or to paint. He remained there until 1492, when he was seventeen, and when this period of beneficent nurturance ended with Lorenzo's death—an event that was far more important to Europeans that year than the Columbian expedition to America. The Medicis were driven out of Florence soon thereafter (but only for a time).

Michelangelo remained in Florence briefly, studying anatomy by performing dissections of dead bodies provided by the Prior of San Spirito Church. These anatomical studies provided him with a fundamental knowledge similar to Leonardo's. Although, unlike Leonardo, he did not leave a large body of detailed "scientific" studies to posterity, some have survived, and his knowledge of anatomy is apparent in all his sculptures, and in his paintings as well.

Eventually he left Florence, traveling and sculpting throughout northern Italy, finally arriving in Rome when in his early twenties. There he was given a commission by a French cardinal to carve a *Pietà* (a statue that shows Mary with the dead Jesus). The result, completed in 1499 when Michelangelo was only twenty-four, may be his greatest work. Mary is depicted as a young woman, because Michelangelo wanted to portray her childlike innocence and virginity. Her face is inexpressibly sad and sweet, but also resigned. The drapery of her gown is a sculptural tour de force. The body of the dead Christ lies on her lap, small in comparison with Mary's massive gown. This was Michelangelo's way of depicting the fragility and transiency of human life, through which the Son of God courageously lived. Michelangelo's anatomical studies clearly taught him a great deal, for the muscles, veins, and skin are perfectly rendered. The left hand of Mary points down, with a single finger extended. How many sculptors would dare to take the risk of creating that finger with a hammer and chisel and not breaking it? Many of Michelangelo's later works were unfinished or partially finished, but this statue is so highly polished that it gleams. Michelangelo was a spiritual man throughout his life, and this work is visibly a labor of inspired religious love. Unfortunately, the statue was attacked and partially defaced in 1972, including the breakage of Mary's finger. It is now restored, but made more remote by a protective glass barrier. I am grateful that I was able to see it several times before the attack, when one could still come close, walk around it, and marvel at it.

Many would say that Michelangelo's next project even surpassed the *Pietà.* It is certainly better known, almost having become his signature work. *David* was produced over a three-year period and was completed when Michelangelo was twenty-nine.

After completing the *Pietà,* Michelangelo returned to Florence, where he was given a special commission. The Florentine Republic had been restored, and the city wanted a statue to honor this return to a

Michelangelo's *The Pieta*

democratic form of government. Michelangelo was chosen to sculpt it. To create the statue, he was given a block of marble that was nineteen feet long, but damaged, so that it seemed too narrow to contain any kind of human figure. His challenge was to somehow find the figure of David inside that piece of marble. The result is the beautiful, compactly muscular adolescent boy, standing gracefully with his weight on his right leg, right arm at his side and left raised to the shoulder, looking intently at Goliath, preparing to slay him using only a sling. The statue represents the glory of the Republic of Florence, ready to overcome all enemies. When moved to the front of the Palazzo Vecchio, the seat of the government, it made a clear statement to the world about the power of Florentine democracy. Towering over the square in front of the Palazzo Vecchio, the statue was dubbed Il Gigante (The Giant).

Throughout his life, Michelangelo thought of himself principally as a sculptor, and he continued to wield a hammer and chisel until his death at age eighty-nine. Many other magnificent works followed *David.* But like Leonardo, Michelangelo was a Renaissance man, a universal genius, a polymath.

He is equally admired for his painting, which flourished during his subsequent years in Rome, as he was given commissions from the Vatican. The most famous are the ceiling fresco of the Creation in the Sistine Chapel and the great fresco of the Last Judgment, painted there on the wall behind the altar at a much later date. Michelangelo labored on the Creation for four years, beginning in 1508, spending nearly every day lying flat on his back on scaffolding rising many feet off the floor. If seeing David inside that peculiar block of marble was a mental feat, what about conceptualizing the entire enormous ceiling while limited to painting only small parts each day by the demanding fresco technique? In both instances the conception was inspired, but the realization required sheer hard work. The Sistine ceiling is for many the ultimate symbol of the act of divine creation. God literally reaches out to an awakening Adam, sending life into his outstretched finger through His own divine creative power.

Sculptor and painter, Michelangelo was also an architect, engineer, and poet. During his earlier career in Florence he designed the New Sacristy in the Church of San Lorenzo, a glorious domed structure containing more of his wonderful sculptures: Night and Day, and Dawn and Dusk, which guard the tombs of two lesser-known Medicis. He also designed and built the Laurentian Library and the grand staircase that approaches it. He designed military fortifications, in this case to protect the city of Florence, using his architectural and engineering skills. Later, in Rome, he redesigned Saint Peter's and gave us the magnificent structure that stands today, personally supervising its construction and refusing to accept any payment for this effort because it was dedicated to God. It was he who conceived of the structure with its two loggias, reaching out as semicircular arms to welcome and embrace the children of God, who would come to worship under a magnificent dome that arched up to the heavens.

Finally, he composed wonderful poetry. One of my favorites is his struggle with the forces of darkness, called *Night:*

> I hug my sleep, and in blocklike rock rejoice,
> Insensible of time's ignominies and injustices.
> Blind, numb, I win; these are my fastnesses.
> O, never rouse me with your ringing voice!

Or there is his description of how he conceives of his mentally created figures, living in a block of stone, in Sonnet 8:

> The best of artists hath no thought to show
> Which the rough stone in its superfluous shell
> Does not include: to break the marble spell
> Is all the hand that serves the brain can do.

What Kind of Environment Nurtures Creativity?

Like Leonardo and Michelangelo, most of the fifty other artists described in Vasari's *Lives* came from noncreative origins. What caused such great creative genius to emerge? What permitted so many diverse creative abilities to flourish? The nurturance that permitted them to become creative did not occur within their family environments. It came from somewhere else. Michelangelo and Leonardo appear to have been born with a "creative nature." However, it might never have become fully manifest if there were no "nurture" to develop it. What were the forces that surrounded Leonardo and Michelangelo—and the other Renaissance artists—that may have helped produce this great flowering of the creative human brain, mind, and spirit? I believe that five circumstances must be present to produce a cultural environment that nurtures creativity.

Freedom, Novelty, and a Sense of Being at the Edge

First and foremost, the Renaissance was an exciting time. After years of medieval scholasticism, the great ideas and techniques of classicism—themselves born in the exciting environment of fifth- and fourth-century (B.C.) Athens—were being rediscovered and reborn.

Until the Renaissance, artists simply copied what their masters had done, and philosophers elaborated on the texts of the Church Fathers. The spirit of the Renaissance is the spirit of breaking out of old and oppressive boundaries, doing what people have not yet done, thinking new thoughts, finding new ways to express, experimenting with new techniques, and exploring new ways to perceive man, nature, and religion. The essence of this kind of environment is intellectual freedom.

And encouraging intellectual freedom is one of the best ways to create creative brains. We have seen in earlier chapters that the creative personality is adventurous, exploratory, tolerant of ambiguity, and intolerant of boundaries and limits. The creative process arises from the ferment of ideas in the brain, turning and colliding until something new emerges. At the neural level associations begin to form where they did not previously exist, and some of these associations are perilously novel. An environment full of intellectual richness and freedom is the ideal one in which to create the creative brain. Renaissance Florence was a bubbling cauldron of such richness and freedom. The works of Plato and Aristotle were being rediscovered and discussed at the Florentine Academy. Polymaths such as Pico della Mirandola were declaring, "I take all knowledge to be my province." Classical art, glorifying the human form and by implication the human mind and brain, was now available to study and universally recognized as a source of inspiration.

A Critical Mass of Creative People

It is more difficult for the creative brain to prosper in isolation. Solitude is usually necessary, of course, for the actual creative process to lead to a creative product. But the catalytic substrate for that process is often interaction with others and intellectual exchange of ideas. Art in Florence was created in studios or shops run by master artists and populated by multiple talented young men. Everyone examined what others were doing. They looked within their own shops, but also at the work of others. They looked at the new creations of their contemporaries, getting ideas about the development of new techniques. They also studied what others of an earlier generation had done. In art alone the city was filled with men of genius, bouncing ideas back and forth and borrowing what was best. Add to that the philosophers, poets, and politicians—it was an astonishingly rich congregation of human beings, who created social networks that cross-fertilized one another and opened avenues from which new ideas could emerge. Another self-organizing system, so to speak.

Put simply, creative people are likely to be more productive and more original if surrounded by other creative people. This too produces an environment in which the creative brain is stimulated to form novel connections and novel ideas.

A Competitive Atmosphere That Is Free and Fair

Much Renaissance art was commissioned by local guilds or other city authorities. Not unlike the present-day competition to design and build the World Trade Center replacement, authorities invited artists or architects to submit designs, and the one deemed to be superior was selected. This "fair freedom" in the economic sphere gave an additional competitive edge to the enhancement of creativity. As we have seen, creative people are individualistic and confident. They may thrive best when pitted against one another.

In addition to such overt competition, Renaissance Florence fostered many other more subtle forms of competition. Art at that time had a variety of set themes, many of them drawn from religion: the Crucifixion, the Last Supper, the Pietà, the Madonna and Child, David and Goliath. The existence of such set themes provided artists with a structure within which they could develop new approaches. The existence of some structure is actually a resource that enhances creativity, since it serves as a reference standard in which new variations on themes can be elaborated. Examples of such structures in other domains are the sonnet, the symphony, the opera, the comedy or tragedy, and the epic poem. The use of set themes also provided another context in which competition could occur. When Michelangelo created his *David,* he and his contemporaries were more than aware that his would be "another David," referring back to those created by other Renaissance masters such as Donatello. His is titanically grander, and no doubt became so because he could demonstrate that he was an even better sculptor than one of the greatest, in both conceptualization and execution.

Mentors and Patrons

Although the creative personality tends to be independent and individualistic, creative people are helped by direct nurturance and support. The mentoring system, whereby a senior expert takes a younger novice under his or her (but usually his for many centuries) wing for training in the necessary intellectual and even social tools and skills needed to succeed, is a fundamental teaching method in modern high-level science. But mentoring has a long history. It derives from a tradition of apprenticeship in many different fields that extends far back in time. The artistic shops and schools of Florence are among the most impressive examples of the power of mentoring.

Mentoring is itself an art. On the one hand, it requires doing teaching and training—providing structure just as specific art forms and set pieces provide structure. On the other hand, a good mentor must also be able to recognize and reward a student whose abilities are even greater. Some claim that Vasari's story about Verrocchio abandoning his paintbrushes when faced with Michelangelo's angel is apocryphal. Even if so, it is a great story with a great moral. As the saying goes, it is a poor teacher who is not surpassed by his students.

Patrons, wealthy individuals who support artists and scientists, may also be important contributors to creativity. Lorenzo the Magnificent was one of the greatest patrons of all time, finding and embracing talented young people and even bringing them into his household. He gave them psychological support as well as financial support, and he also was instrumental in producing a "critical mass" in the Florentine creative environment. As his accomplishments illustrate, the role of a patron is not simply limited to financial support. The patron gives the artist or scientist an important vote of confidence from a prominent and respected person. Although creative people are confident even to the point of arrogance, they are also self-critical and perfectionistic, and these latter forces may inhibit their creativity. The emotional and intellectual support of a patron is an important nurturing resource that counters those inhibitory forces.

Economic Prosperity

Almost all periods of great creativity, populated by many creators, have been times of economic prosperity. This may not be a mandatory component of a creativity-enhancing environment, but it is certainly helpful.

Economic prosperity feeds creativity in several ways. It provides the accumulation of intellectual resources in which ideas can be stimulated, in which they can bubble and ferment—the collections of earlier art, the libraries, the salons, and the gardens where people can meet and discuss and argue. It provides the financial resources to attract the numerous people who form the critical mass. It provides the funds for their raw materials—the marble, the paint, the paper, the wood, the glass. It provides the funds to commission artists and pay them for their work. As great cities form and remodel themselves architecturally because of economic prosperity, a visual atmosphere is created that is itself inspiring.

The Importance of Environment

These five factors also characterize the other "cradles of creativity." An atmosphere of intellectual freedom, ferment, and excitement was also prominent in fifth- and fourth-century Athens, nineteenth-century Paris, late nineteenth- and early twentieth-century America, the Enlightenment, Tudor and Elizabethan England, and Revolutionary America. So was a critical mass of creative minds, free and fair competition, mentors and patrons, and at least some economic prosperity. If we seek to find social and cultural environmental factors that help to create the creative brain, these must be considered to be important ones.

The environment into which an individual is born makes a difference. Had Leonardo or Michelangelo been born two hundred years earlier or later, we would never have had the body of work that they produced. Anatomical dissections would not have been possible at an earlier time. Patrons and prosperity would not have been there to support them. Without Lorenzo, Michelangelo would not have been a sculptor. Had Julius II not commissioned the Sistine ceiling, Michelangelo would not have turned his hand to fresco. Both Leonardo and Michelangelo would have had a "creative nature," but it might never have become manifest had they lacked the nurture of a supportive environment. So too it could be for other great creators—Phidias, Plato, Aristotle, Monet, van Gogh, William James, or the Wright Brothers. As I sometimes say (though I hardly rank with the geniuses I have been describing), if I had been born one hundred years ago, I would never have been a neuroscientist or a physician. Neuroscience did not exist, and only rarely were women allowed to study medicine, or even to attend college.

Environment makes a difference!

The Role of Nature: Innate Gifts and Hereditary Factors

But so does nature. "Nature" is related to heredity, but not identical.

Where does creative genius come from? How does it arise? These are both questions that everyone would like to answer.

Our case studies also shed light on these questions. Think about how the brains of Leonardo and Michelangelo were created.

When these two men were conceived, they were the consequence of shuffling the cards in the genetic deck, with half of the genes coming from the mother and half from the father. A grand challenge in modern science is to figure out how those shuffled genes become translated into complete living organisms, many of them complicated almost beyond imagination. A human being, for example, is somehow produced by forty-six chromosomes and about thirty thousand genes, give or take a few. Somehow these genes must orchestrate the creation of cells and their differentiation so that they form diverse body organs, such as the liver, the kidneys, and the brain. Within the brain there are also many different types of cells. But, most importantly, the human brain is defined by the multiple and complex ways that these cells are connected to one another. At this moment we still know very little about how genes affect the development of the brain in the uterus before birth, or during childhood, adolescence, and adult life. We are almost totally clueless about how genes become translated into complex human traits such as creativity or personality or cognitive style. Lots of people are studying the genetic regulation of small parts of the process. We know that genes produce proteins with funny names like MAP or GAP or SNAP, which affect components of brain development and maturation, such as neurogenesis or synapse formation. But we do not know much (yet) about how genes affect the interconnectedness of the trillions of neurons in our brains and the quadrillions of synapses that talk back and forth to one another. Thus we can say nothing at present about how genes, working at the molecular level, might have an influence on the creation of the creative brain. For now, we have to rely on speculations and hunches, combined with crude empirical methods, such as family studies of heritability.

What we perhaps can say is that Mother Nature gives creative people brains that are well designed for perceiving and thinking in original ways. Some of that influence must be coded in the genetic shuffle in ways that we do not yet understand. And very likely the gift given by Mother Nature is an enriched ability to make novel associations and to self-organize in the midst of apparent disorganization or even chaos.

The creative brain may appear unexpectedly, in people who simply seem to have been given innate gifts. Or it may appear within a hereditary context, in people who seem to have a genetic endowment that makes them creative.

Innate Gifts

One fact that we have to reckon with is that many creative people are creative "by nature," without any obvious evidence that their creativity is due to genetic factors. These are people whose creativity came "from nowhere" in the genetic sense, but who appear to have had innate gifts even during early childhood. Within them they have a creative drive and passion that cannot be suppressed. How does this occur? At present no one knows. Perhaps distant forebears whom we do not know about were creative, and they provided some creative predisposition. Perhaps these ancestors provided a sufficiently additive accumulation of ordinary creativity to hit a home-run combination in the genetic shuffle. Perhaps an incredibly rich environment made a difference. Perhaps, as Vasari might argue, the gift came directly from God.

In the case of Leonardo and Michelangelo, we know only about their immediate family history. In both cases the fathers came from the lower ranks of the merchant class and were struggling to maintain their status (although both Michelangelo's parents had distant connections with the nobility, for whatever that is worth). Leonardo's genetic deck shuffled his merchant father with an illiterate peasant girl. Michelangelo's was shuffled through two parents with similar social background. Nothing that we know suggests any family history of creative contributions for either man on either side of the family. However, being from the merchant class may require more spunk or drive than being from the leisure class.

In early childhood both were noticeably and precociously talented. Leonardo drew, and he sang using the lute—and did both very well. Michelangelo excelled at drawing. Both had brains that were innately gifted with an ability to observe the world around them and reproduce the images created in their minds by transforming them onto paper, without ever being taught to do so. It was in their nature. In terms of formal education, neither had much. Did this actually enhance their creative capacities, by freeing them from preconceptions about the world or from rigid rules and structures? Possibly. Did growing up in a more rural area help them to become more creative, by giving them a greater spatial extent of land and sky and a greater diversity of animal life to study than they might have had as city dwellers? Possibly.

At an early age they must have been honing their mental capacities to perceive and record three-dimensional relationships, simply by seeing and thinking and making mental manipulations of their observations. On this base they would later build their interests in anatomical dissection and rendering the human body accurately, their ability to solve architectural and engineering problems that eluded others, and of course their skills as painters and sculptors. Somehow an innate gift, intrinsically coded in their brains by genetic influences that we do not as yet understand, was present. It is manifested by cognitive and personality traits such as curiosity, openness to experience, and self-confidence. These traits can be further enhanced by environmental influences, and probably were in both men, because the human brain is "plastic." That is, it is shaped intensively throughout life by interactions with the world around it.

Then, in early adolescence, both were identified as potentially gifted artists. Both became apprentices to great artistic masters of the time. Both quickly learned the "basics" and soon surpassed their masters. In short, they were launched on the careers that they would follow for the remainder of their lives. Each apparently realized, at a very young age, that he was a genius. Each was driven to be original, independent, and creative. Given that both were fortunate enough to be in Renaissance Florence in the fifteenth century, nothing could have stopped either of them, apart from physical injury or disease. Fortunately, neither had either. Each continued to develop, to modify and to use his brain, in slightly different ways, depending on the political and social forces surrounding him.

In terms of heredity, many other prototypical geniuses we have seen in these pages are not dissimilar, in the sense that they had innate gifts. They were born with a "creative nature" into families that were not particularly creative or highly educated. Think of Shakespeare, with his merchant father and his "small Latin and less Greek." Or Newton, Einstein, Ben Jonson, Ben Franklin, Picasso, or (more humbly) even Lewis Terman.

Hereditary Factors

A "creative nature" can, however, also run in families. In this instance, we are prompted to conclude that creativity may be due in part to hereditary factors—by which we actually mean genetic influences that are definitely present, even though we do not understand how they are working at the molecular or cellular level in the brain.

Our largest source of evidence for the heritability of creativity is anecdotal accounts. There are a variety of famous families in which at least two members have made significant creative contributions. I have already mentioned the Darwin/Galton family: grandfather Erasmus Darwin was moderately creative, and Charles Darwin and Francis Galton were highly creative. In *Hereditary Genius,* Galton himself summarized multiple pedigrees that contain at least two creative people. He divided them into various fields, some of higher creative impact than others: judges, statesmen, poets, painters, scientists, musicians, and others. For comparison he included two types of athletes: oarsmen and wrestlers. *Hereditary Genius* is not perfect, but it is a highly informative compendium of pedigrees of gifted families. Even today, almost 150 years after publication, it still makes interesting reading.

The Bach family, for example, is perhaps the most powerful example of creativity running in families. Its creative members extend over eight generations, beginning in 1550 and ending in 1800. The greatest was, of course, Johann Sebastian Bach. But in addition to him, there were more than twenty eminent musicians in the Bach family. Other families summarized in Galton's *Hereditary Genius* include the Bellinis and Van Eycks and Titians among the painters, the Coleridges and Wordsworths among poets, and the Brontës among novelists. Galton described many other examples, but these are some of the best known.

It is not difficult to find gifted families from more recent times that might be included in an updated edition of *Hereditary Genius* as evidence for the familiality of creativity. For example, Thomas Henry Huxley ("Darwin's bulldog") was a notable scientist who had three distinguished grandsons. Grandson Julian was an anthropologist who carried forward his grandfather's work on the theory of evolution. Andrew was a physiologist who received the Nobel Prize in Physiology or Medicine for his work on nerve impulses and muscle contraction. Aldous was the author of *Brave New World* (1932) and *Point Counter Point*

(1935) as well as numerous other writings. The brothers William and Henry James are another example. Their father was a nineteenth-century American intellectual who was a close friend of Thoreau and Emerson. William became a distinguished philosopher and psychologist. Henry became an equally famous novelist.

From a scientific perspective, however, such anecdotal accounts have significant limitations. Families are selected because they provide positive evidence for heritability. But anecdotal accounts do not tell us how often there is *no* evidence for heritability.

To my knowledge, my Writers' Workshop study is the only piece of creativity research that has made an effort to examine heritability using a well-selected sample of creative people and a comparison or control group. In addition to determining the extent to which the writers suffered from mental illnesses, I also examined the extent to which their parents, brothers, sisters, and children had mental illnesses, and the extent to which they were creative. The results are very interesting, and they can be interpreted as providing partial support for a hereditary contribution to both mental illness and creativity.

Following a classification scheme that had been developed by Swedish-American psychologist Tom McNeil for his studies of the heredity of mental illness and creativity, I divided the writers' relatives into three groups: not creative, moderately creative (+creative), and highly creative (++creative). I considered them +creative if they pursued occupations that are somewhat creative, such as journalism or teaching music or dance. I considered them ++creative if they had a well-recognized level of creative achievement, such as writing novels, performing as a concert artist or playing in a major symphony, dancing in a major company, or making a major scientific contribution such as an invention.

The 30 writers had a total of 116 relatives, and the 30 controls a total of 121. Among the writers, 32 relatives (28%) were creative (20 +creative and 12 ++creative). Among the controls, 16 relatives (13%) were creative (11 +creative and 5 ++creative). Statistical tests showed that these differences were significant (not likely to be a chance result).

TABLE 5-1. Patterning of Mental Illness and Creativity in 30 Writers, 30 Control Subjects, and Their Families

	Writer			Writer's Family		Control Subject		Control Subject's Family	
Subject Number	Affective Disorder	Bipolarity	Alcoholism	Mental Illness	Creativity	Affective Disorder	Alcoholism	Mental Illness	Creativity
1	●	●		●	●				
2	●		●	●	●				
3				●					
4	●	●	●		●				●
5	●	●		●		●			
6	●			●					
7		●		●	●	●			
8	●	●	●	●	●			●	
9	●	●	●	●	●			●	
10	●	●	●		●				
11	●	●		●			●		●
12	●	●		●	●				●
13	●				●				
14	●			●				●	
15	●		●	●		●			●
16	●			●	●				
17	●		●						
18	●					●			
19	●								
20	●				●				
21	●	●	●	●	●	●			
22	●					●	●	●	
23	●			●	●				
24	●								
25						●		●	●
26	●	●						●	
27				●					
28						●			●
29	●		●	●	●				
30	●					●		●	

TABLE 5-2. Iowa Writers' Workshop Study: Psychiatric Illness in First Degree Relatives of Writers and Controls

	Writers, N = 30		Controls, N = 30	
	N	%	N	%
Parents				
Total Parent N	60		60	
Bipolar	1	2	0	0
Unipolar	9	15	1	2
Any Mood Disorder	10	17	1	2
Alcoholism	5	5	4	7
Suicide	2	3	0	0
Any Illness	25	42	5	8

	Writers, N = 15		Controls, N = 15	
	N	%	N	%
Siblings				
Total Sibling N	56		61	
Bipolar	3	5	0	0
Unipolar	8	14	2	3
Any Mood Disorder	11	20	2	3
Alcoholism	3	5	3	5
Suicide	1	2	0	0
Any Illness	24	49	5	8

The larger number of ++creative relatives among the writers is especially noteworthy. The relatives of writers also had a higher rate of mood disorder than the controls, 18 percent in relatives of writers and only two percent in relatives of controls. So it appears that both mood disorder and creativity were familially transmitted.

When summarized in a numerical list that shows the patterns of creativity and illness in the writers, controls, and families, the difference between writers and controls is striking. This can be shown in a patterning chart. Only nine of the writers had family backgrounds that had neither creativity nor mental illness, whereas the majority of the controls (18) were free of a hereditary association with either trait. These results suggest that creativity and mood disorder are indeed closely linked in writers and their families. Something seems to be transmitted that may predispose descendants to both characteristics.

However, a smart reader is going to notice that this study does not show that creativity is hereditary, in the sense that it is directly genetically transmitted. It only shows that it runs in families. It could run in families because of genes, but it is also very possible that it is a learned trait that occurs because of growing up with other family members who are creative. Think about those hundreds of musicians in the Bach family. They were definitely taught to be musical by growing up in a musical family. In fact, the whole Bach extended family would get together every year and have family concerts. Likewise, literary people may teach their children writing skills at an early age and predispose them to become writers. Dads who were successful football or soccer players in high school or college are usually throwing or kicking balls with their children as soon as they are physically able. My study might only be showing an increased rate of creativity in the families of writers because they shared an enriched environment, not because creativity may be inherited through a genetic mechanism.

But some clues in my study hint that the transmission of creativity might be at least partially genetic. Most noteworthy is that the types of creativity that were found in writers' family members were not necessarily literary. While some were in literary fields, many were creative in other areas, such as art, music, dance, mathematics, or science. This variability in types of creativity argues for a genetic role. If the creativity were a socially learned trait, one would expect that the majority of the creative family members would be writers as well. But they were not. The coexistence of multiple forms of creativity within these families suggests some kind of "general creativity factor" that is genetically transmitted and that predisposes a person to be original and inventive. The individual who gets a goodly dose of this factor can then differentiate into a composer, a violinist, a dancer, a mathematician, or a biologist, depending on physical abilities, encouragement from family or teachers, and a host of other physical or environmental influences.

Having said that, I should make it clear that my study has only a limited ability to inform the nature versus nurture debate. This study, like all studies using a design that examines only familial patterning, cannot definitively disentangle genetic from environmental factors. Only one experimental design can do that, the adopted offspring study. That design cleanly separates genetic from environmental factors by studying children who were adopted away to "normal" or "average" families, but whose biological mothers possess some trait or disease of interest, such as schizophrenia or creativity. Those adopted children can be compared to another group of adopted children who were born to biological mothers who were "normal" or did not possess the trait of interest. If the children adopted from schizophrenic or creative mothers have a higher rate of schizophrenia or creativity than those adopted from normal mothers, then we can say that the transmission of the trait has a significant genetic component. The influence of the environment—being raised by a mother who has schizophrenia or who is creative—has been eliminated.

This design has never been used to study the hereditary transmission of creativity, and it probably never will be, because it is fiendishly difficult to execute. Interestingly, Tom McNeil did do a variant of it by starting with a group of adopted children who were divided into three groups with different levels of creativity. These were essentially equivalent to my "absent," "+creative," and "++creative." He too was interested in the relationship between creativity and mental illness. Because the study was done in Denmark, where health registries designating diagnoses are maintained in a relatively open and accessible way, he was able to determine the extent to which creativity in these adopted children was associated with higher rates of mental illness in biological parents. What did he find? No big surprise. The more creative the adopted person, the higher the rate of mental illness in the biological parents. Unfortunately, he had no way to find out about creativity in the biological parents. If he had, we would have a clean answer to our question about the degree to which genetic factors directly influence levels of creativity.

Nature versus Nurture: What Creates the Creative Brain?

As with so many of the most interesting questions in science, we have (as yet) no definitive answers. But we can make some reasonable observations that will guide us as we continue to seek answers.

First, we need to conceptualize "nature" in two different ways.

"Nature" can be defined as an innate or inborn gift that drives an individual to creative achievement, without any obvious genetic contributions. We do not know yet how this kind of creative nature arises, but it appears to be more common than "nature" that is clearly hereditary. Once this creative nature arises, nurturing it through a variety of environmental factors will further enhance it. If it arises where there is no nurture at all—as might occur during dark times of the Middle Ages or ferocious times and places wracked by civil war or political or religious suppression—the precious innate gifts may be extinguished and fail to produce any great art or science at all.

"Nature" can also be defined as a predisposition to be creative that appears to be hereditary. Some evidence supports the possibility that a tendency toward creativity may be inherited. But this evidence is not definitive, nor is it likely to become so anytime soon. At present the evidence is primarily anecdotal. We need more well-designed empirical scientific studies of the heritability of genius.

Second, whatever the importance of "nature," "nurture" is also important for creativity to flourish, and perhaps essential. The human brain is shaped by the world around it from the time that a child is born to the end of adult life.

Many different forces have an impact on how the brain grows and develops. One of our challenges is to understand those forces more deeply, so that we may eventually use that understanding more wisely. Thus we may give the gifted an opportunity to shine yet more brilliantly. And we may also help the more ordinary build better brains as well.

Which brings us, finally, to us.

6

BUILDING BETTER BRAINS

Creativity and Brain Plasticity

What a piece of work is a man! how noble in reason! how infinite in faculties! in form and moving how express and admirable! in action how like an angel! in apprehension how like a god! the beauty of the world, the paragon of animals!

—William Shakespeare, *Hamlet*

If Shakespeare were writing those lines in the twenty-first century, he might instead say, "What a piece of work is the *human brain!* how noble in reason! how infinite in faculties!"

The growth and development of the human brain is in fact even more miraculous and awe-inspiring than those lines from *Hamlet* might suggest. If anyone could come up with the appropriate level of eloquence to describe it, the Bard of Avon would be the person. But even he might be tongue-tied and pen-tied (and now keyboard-tied) when confronted with the wondrous process of brain development.

If we want to understand how to build better brains, we need to understand how the brain develops and continues to grow and change throughout life. The process begins during the second or third month after conception. Development prior to birth is primarily guided by the "genetic blueprint" produced by the intermingling of the 23 chromosomes received from the mother and the 23 from the father. Each person has a unique genetic blueprint. To some extent this genetic uniqueness dictates the individuality of our minds and brains—whether we can taste particular flavors, whether we have an aptitude for mathematics or poetry, or perhaps whether we have an innate capacity to think creatively.

This genetically preordained blueprint lays out a general plan for how the brain will develop throughout pregnancy. As a first step, nerve cells form along a small neural plate in the middle of an organ that is so primitive that it can scarcely be called a brain. The cells then slowly begin to migrate outward, gradually lining up to form the cerebral cortex. As I mentioned in chapter 3, there are approximately 100 billion neurons in the cerebral cortex. It is awe-inspiring to contemplate the complexity of producing these 100 billion neurons and then sending them out on their pathfinding mission. The Yale neuroscientist Pasko Rakic was the first to discover that they are able to achieve the miraculous task known as *neuronal migration* because they "ride a glial monorail" to reach the correct destination, following a path piloted in advance by guide cells known as *glia* (the Greek word for "glue"). With this assistance, the 100 billion neurons somehow manage to arrive at the correct destination and eventually line up in six orderly layers.

As the neurons arrive in the cortex, the brain also begins to "wire" itself, through processes known in neuroscience as *axon formation, dendritic proliferation,* and *synaptogenesis.* Creating the correct wiring is another awe-inspiring task. The complex set of networks that permits the human brain to be the most elegant self-organizing system in the universe is beginning to take shape while the infant is still growing in the mother's womb. The newly arrived neurons begin to form connections, both in their nearby microscopic neighborhood and in more distant regions that work on a larger and more macroscopic scale. Each of the 100 billion neurons must "reach out and touch" approximately 1,000 target cells by sending out axons (the "wires" of the brain), using delicate sensors known as *growth cones,* which form on the tips of the axons, to make contact with the target. We now know, through the work of the Berkeley neuroscientist Corey Goodman and others, that the axons' stretching and touching is guided by a group of chemical substances that produce forces of attraction and repulsion, ensuring that the axons arrive at the correct location and that things are hooked up correctly. Dendrites begin to sprout, and synapses begin to form. As this process proceeds, the brain grows and expands in size. By the seventh month in the womb, it begins to resemble a real human brain.

Depending on where the neurons arrive and how they become wired to one another, they also differentiate to create specific functional regions. The functional regions develop so that after birth the child can begin to learn to perform a variety of necessary tasks: seeing, hearing, smelling, moving the body, and all the other things that our brains command us to do.

The brain maturation that begins in the fetus is still at a very primitive level at the time of birth, however, as evidenced by the nearly complete helplessness of the human infant. Human brain maturation is a very long process. It continues after birth and persists even through adolescence and early adulthood. During this time, the brain continues to wire itself, with each neuron producing more and more synapses to connect nearby and distant brain regions. If you ever have wondered what the process is that permits a newborn child to learn to crawl, walk, and talk, it is the ongoing process of "brain wiring." During childhood and early adolescence, the connections actually "overgrow." Much as a gardener sows many seeds and then selects only the most hardy young plants, weeding away the rest, so the brain creates more connections—more spines and synapses—than it actually needs. During late adolescence and young adulthood, a process known as "pruning" occurs, and the overgrowth is trimmed back so that the brain can work effectively, efficiently, and maturely. If you have wondered what it is that makes your somewhat obnoxious teenager turn into a pleasant adult, it is probably because his or her supercharged but unpruned brain has finally gotten a good trimming.

When the newborn child emerges in the oxygen-breathing world and lets out its first cry, its brain faces a totally new set of challenges. The genetic blueprint that governed brain development so effectively in the womb must now share its governance with a host of nongenetic forces. Nature must bow its head in the face of nurture. *Brain plasticity* is the process by which this happens. It is a critical concept: it is a key tool for building better brains.

What Is Brain Plasticity?

When we neuroscientists say that the brain is "plastic," we are not talking about polymers. We mean that the brain is marvelously responsive, adaptable, and eternally changing. Its adaptations and changes occur in response to the demands and pressures of the environment that it encounters. Sigmund Freud and the psychoanalytic movement gave us an awareness that early life experiences affect emotional development and attitudes in later life. Neuroscience adds a new dimension: it makes us aware that experiences *throughout* life change the brain throughout life. We are literally remaking our brains—who we are and how we think, with all our actions, reactions, perceptions, postures, and positions—every minute of the day and every day of the week and every month and year of our entire lives. Your brain is being changed by the process of reading this book. (For the better, I hope.)

During infancy, childhood, adolescence, young adulthood, middle age, and late life we all accumulate a trove of experiences and memories. These shape our minds and brains, and mightily so. We literally become what we have seen, heard, smelled, touched, done, read, and remembered. Some of us have smelled cookies freshly baking and have tuned our brains to feel both soothed and hungry at that scent. Some of us were given a violin or piano to play. We have been scolded and praised in many different ways, providing us with a learned value structure. Those of us who grew up in "radio days" have different memories, and probably differing auditory and imaginative skills, from those of us exposed to the graphic visual images that flicker across a television screen. Those of us who grew up doing arithmetic "in our heads" in the precalculator and precomputer era may have greater skills at doing various mental manipulations, yet we watch with awe and even envy as five-year-olds swiftly navigate their way through icon-driven menus on any one of the myriad handheld or desktop computerized devices that currently surround us. Differences in the environment to which their brains have been exposed have produced very different brains from those of a sixty-year-old.

What are the mechanisms by which the plastic brain continually remakes itself?

Ten or fifteen years ago it would have been impossible to answer that question. Now, however, neuroscientists have worked out explanations on several different levels. Much of this is highly technical, so I'll just provide a very simple overview to give you an appreciation of how much we have learned about the brain's amazing capacity to learn, change, and adapt. This occurs simultaneously on the fine-scaled level of molecules and on the large-scale level of brain systems.

One important part of brain plasticity is the ability to retain and store specific memories. For many years people who are interested in how the brain works have pondered a very fundamental question: how are memories stored? The storage of experience as memories is the foundation upon which the brain builds its capacity to continually remodel itself. The mechanism for memory storage has been elucidated during recent years principally through the work of psychiatrist and Nobel Laureate Eric Kandel. As it turns out, memory storage is carried out on the molecular level, and the synapse is the workhorse for ensuring that memories are preserved for long periods of time. Even a simple description of how this happens will provide an enlightening example of how complicated human brain biology actually is.

Preservation of memories over the short term occurs because existing synapses are strengthened. Long-term memory storage must be produced by the creation of new synapses and even enlargement of the dendritic arbor. To build something new, the command center of the cell, the nucleus, must be drawn into the picture. The nucleus stores genes. And the genes contain the information that regulates growth and change in the cell. But genes only release that information when something stimulates them to act—that is, in the language of genetics and cell biology, when they are *expressed.* In other words, genes themselves are influenced by the environment both inside and outside the cell! In this particular case, when the neuron is stimulated to a sufficient degree to create a memory that needs to be preserved, a variety of chemical messages are sent to the cell nucleus, where in turn genes are expressed and send messages back out to the synapse that say: "build some more synapses and create new synaptic connections so that you can keep this information for a long time." The actual messages are proteins created at the command of genes, which enhance the building process. They have names that you do not want to remember, such as ubiquitin or CREB (Cyclic-AMP Responsive Element Binding protein). (No doubt Eric, his lab team, and a few of his close friends can remember what CREB stands for. I included it because it is actually very important for memory storage, but even I do not want to tie up any of my precious synapses to store the details about the acronym in my long-term memory. I had to look it up.)

In parallel with the remodeling that occurs on the small scale of cells and molecules, plasticity also occurs in the brain on a large scale as well. For example, the brain has specific regions dedicated to performing particular functions, such as moving, touching, or seeing. In the cortex these are recorded as "maps" that reflect the location of the function and the amount of brain tissue that is actually needed to perform it. Perhaps the most famous of these is the *motor homunculus,* a map of where the various motor functions occur on the "motor strip," a band of cortex in the middle of the brain. As a picture of the homunculus indicates, our most important motor functions claim larger shares of cortical real estate. The hand has more space than the foot, for example. Our mouths—so important for talking, eating, and kissing—occupy a large share as well. But we now know too that cortical maps may also change in response to environmental influences. If several fingers

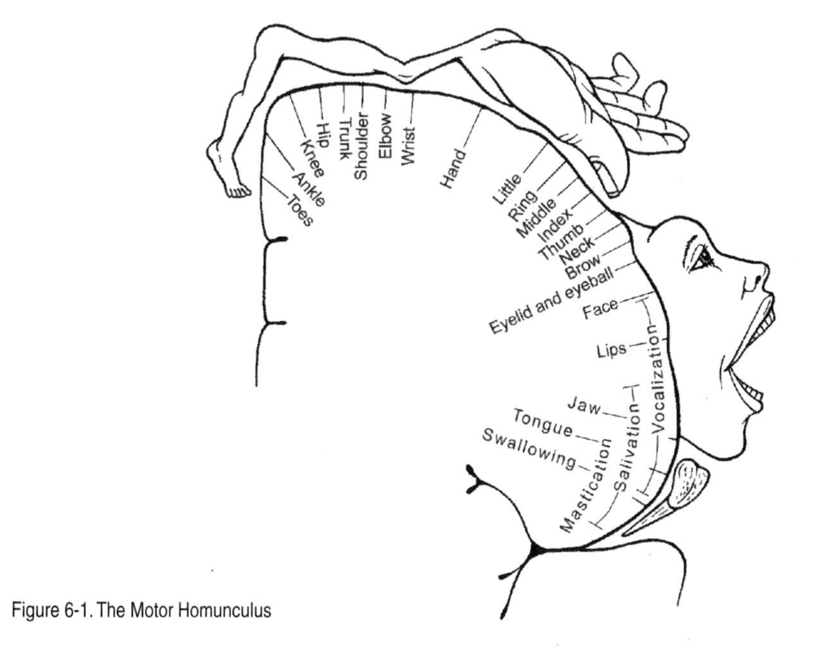

Figure 6-1. The Motor Homunculus

are lost from a hand, for example, their claim to cortical real estate is gone, and the remaining fingers move in to claim it. This is a simple example that is well established from empirical data, but it makes the important point that large-scale remodeling of the brain occurs as well. The brain is plastic and dynamic on all levels.

Another important component of brain plasticity is the concept of *critical periods.* These are relatively limited windows of opportunity during which the brain can learn, change, and develop. If the opportunity is not seized during that critical period, the window may close forever. The concept has many important implications for building better brains, for enhancing creativity, and for education.

The fact that the brain has critical periods in its development was first demonstrated in a group of experiments done by Torsten Wiesel and David Hubel, for which they were awarded a Nobel Prize in 1981. They showed, in both cats and monkeys, that preventing input to one eye during an early stage of brain development produced abnormalities in brain development. Because one eye was unable to see, the visual cortex did not grow properly. The tidy columns of nerve cells normally seen in that brain area failed to line up neatly, with the consequence that the animals were never able to recover the normal columnar structure. Instead of the neat stripes that characterize normal visual cortex, the animals had a hodgepodge. The abnormalities created in the brain produced abnormalities in visual function. The binocular vision that permits us to have depth perception arises from that striped visual cortex, which codes input from two eyes. The animals had forever lost binocular vision and depth perception. No matter how much visual input the eye received after it was unpatched after the critical period, the lost binocular vision could not be recovered.

Examining everyday human life, we can find many examples of lost or diminished capacities that have occurred, or that may occur, because of inadequate environmental exposure and a resultant failure to learn something during a critical period. One obvious and important example is the ability to read, write, and speak foreign languages. Although most other developed countries usually require children to begin studying foreign languages while in grade school, American schools usually do not. Studies of neuroplasticity have made it clear that the prime time for language acquisition is between around age one and around age twelve. This is when children are training their ears and brains to hear subtle differences in sounds and to articulate them with their lips and mouths. It is truly a "no-brainer" that foreign languages should be introduced into our schools during the elementary grades. Unless of course Americans wish to become intellectual isolationists or English-language imperialists! (One of my pet peeves is to be in a foreign country and to observe my fellow Americans complaining that "they don't speak English here." The honor of being a world power carries with it certain responsibilities, and one of them should be to understand other languages and cultures apart from our own.)

Some of us worry that too much early childhood exposure to the vivid visual images shown on television or through the Internet may be shaping the way that a generation of children will learn to perceive the world during a critical phase of their development. This worry has sev-

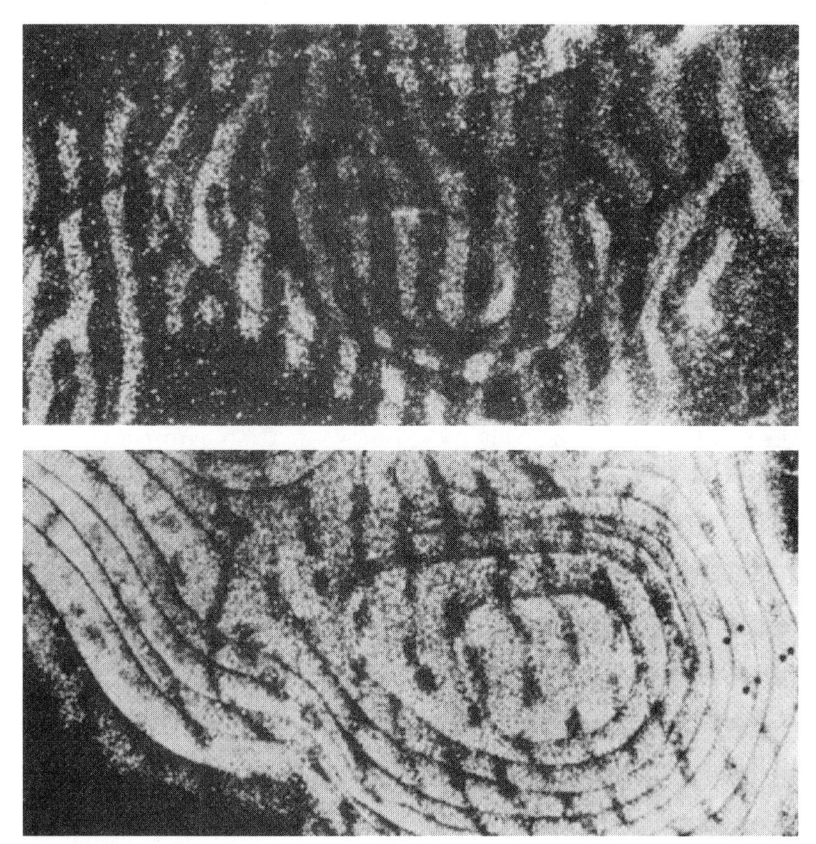

Ocular Dominance Columns. A normal, neatly striped pattern is shown above. Below is the "hodge-podge" that occurs when an eye is covered or lost during a critical period.

eral facets. One is content. Very young children, living in an "average" home that subscribes to fifty to one hundred television channels and is equipped with a remote control, can easily find their way to scenes full of nudity, explicit sexuality, and violence. By the time they reach puberty, their brains will be sexually primed, and their libidos will perhaps be proportionally higher than their intellectual or moral maturity. Young children have seen people shooting other people so often in movies and video games that they are desensitized to what it actually means. How many Columbines will we continue to have?

Another facet of the worry about too much vivid visual exposure during childhood is the potential effect of television and the Internet on creativity and originality. Visual input from TV and film is totally passive, and much Internet material provides passive input as well. A child who reads a book about Robin Hood or Harry Potter is learning to visualize and imagine for herself. If derived from reading, Claire's mental vision of Harry and Hogwarts will be generated actively rather than passively. And it will be different from Owen's. Once Claire and Owen have seen Harry and Hogwarts shown to them in a film, however, the power of that vivid visual film image is likely to replace what they created for themselves. Instead of the many Harrys occurring in the minds of many children, there is now one "standard Harry" that many children will passively adopt rather than creating their own. As Darwin has pointed out, evolution thrives on variation. And creativity does as well.

There are many other somewhat more trivial examples of the importance of critical periods. For example, typing rapidly is a motor skill that can be learned relatively early and that is necessary in our computerized world. Yet many males of age forty or older have never learned to type, because it was once assumed that they would never need to. They would have handmaidens (wives or secretaries) who would perform this supposedly menial chore. Now they are genuinely handicapped as they attempt to answer e-mail or create a summary of a recent meeting while working with their computer on an airplane, slowly hunting for the right letter or number with their eyes and pecking away with two forefingers. Although a motivated person can learn to type as an adult, it is a much slower process than when acquired during childhood or early adolescence. Similarly, sports fans have watched the age of star tennis players, gymnasts, golfers, and swimmers drop to steadily younger levels, because training in these sports now begins sooner than in previous generations, to take advantage of developmental opportunities. These are all examples of capitalizing on acquiring information and skills during critical periods (or failing to capitalize, in the case of diminished capacities). Creating the right environment to learn during the right time to learn is one of the secrets of building better brains.

Plasticity and the Creative Brain

Our discussion of brain plasticity so far has emphasized understanding the neural basis of plasticity. This leads naturally to an important question: Do we know anything about how experience has changed the brain in people who are creative?

Fortunately, the answer to that question is yes.

When I began my studies of creativity in the Iowa Writers' Workshop some thirty years ago, it was impossible actually to study the writers' brains. Their personalities, yes. Their work habits, of course. Their subjective descriptions of the mental processes they used to produce poems and novels, certainly. Their intellectual ability and cognitive style could also be measured. All these things are indirect indicators of how the brain works. But the crucial command center, the brain itself, was inaccessible to direct scientific scrutiny. It remained inscrutable and inaccessible, carefully protected by the hard, rigid, and impenetrable structure of the human skull. For someone like me, who went to medical school hoping to understand the human brain, all this was very frustrating.

Beginning in the late 1970s, however, things began to change. Computerized tomography (CT) was invented by Godfrey Hounsfield and Alan Cormack, netting them a Nobel Prize in 1979. Magnetic resonance (MR) imaging was invented in the 1980s by Paul Lauterbur and Peter Mansfield, for which they won a Nobel in 2003. The neuroimaging revolution was launched. Scientists like myself were given tools that we could use to visualize and measure brain anatomy and physiology in living human beings. MR imaging has been an especially powerful tool, because we can use it to measure brain anatomy via morphometric MR (mMR), brain chemistry with spectroscopic MR (sMR), and indirect indicators of changes in blood flow with functional MR (fMR). I was one of the leaders in applying these wonderful tools to studying the brain, but my major emphasis has been on using them to understand the brain abnormalities that occur in serious mental illnesses, because they are perhaps the most important public health problem in our contemporary world. Faced with all the human misery that these illnesses produce, studying the creative brain seemed like an intriguing luxury that I couldn't afford, at least for a while.

Fortunately, however, others have responded to the opportunities afforded by neuroimaging to understand the creative brain.

We now know that when we store memories we increase the number of synapses in the normal healthy brain. That might actually translate into an increase in the volume of gray matter in the specific regions where the synapses increase. Is it possible that we could use the capacity of mMR imaging to determine if a measurable change in size actually occurs?

One of the first studies to address this question chose an interesting group of subjects who were required to have significantly enhanced spatial navigation skills, albeit not otherwise conspicuously noted for being creative. In 2000 Eleanor Maguire and her colleagues at University College in London studied the size of the hippocampus in London taxi drivers. The hippocampus (Greek for "seahorse," because this structure looks like one) is well known to neuroscientists as a site where memories are consolidated. London taxi drivers were chosen for this study of the effects of navigational skill learning on brain plasticity because they are subjected to intensive training that lasts for two years, which requires them to learn how to locate thousands of sites within the city and to pass strenuous police examinations. Maguire's group found that the taxi drivers had an increase in size in the hippocampus compared to a control group, and that the increase was correlated with the number of years spent driving a taxi. They concluded: "It seems that there is a capacity for local plastic change in the structure of the healthy adult human brain in response to environmental demands."

These assumptions and this approach were subsequently used to study a group of people who are creative: symphony orchestra musicians. Vanessa Sluming and her colleagues at the University of Liverpool capitalized on an interesting aspect of the mental abilities that are required for people who play in symphony orchestras. Most of us would assume that orchestral musicians simply need to be competent in performing on their specific instruments. If we have a bit more sophistication, then we would be aware that orchestral musicians are in fact part of a large group that has to create sound together in a cooperative and interactive way, aided by guidance from the conductor. But few of us might realize that orchestral musicians also need to have highly developed visual/spatial abilities, because orchestral performers have to be skilled at reading printed music and also have to orient their responses and performance to those organized in the space around them.

Reading music is an acquired skill that is similar to reading words, but with a few differences. A primary difference is that the material being read is musical notes, rather than the letters of the alphabet. But the notes are also displayed spatially in a vertical dimension, on a "staff" of horizontal lines that indicate the degree of difference in pitch. To actually "read" the notes, the performer must observe how far apart each note is from the others in pitch (interval analysis) and also follow the notes across the page, heeding cues for duration and tempo. Most skilled musicians are able to hear the notes simultaneously in their heads (i.e., brains) as they read them. (This is different from skilled readers, who generally do not silently pronounce the words that they are reading, because this would slow them down.) In general, individuals who are skilled at reading music begin their training at a relatively early age. Young musicians trained with the Suzuki method, which teaches them to imitate the sounds of music that they hear as a first step during years three through six, usually begin to learn to read printed notes at around age six, in parallel with a time when they are learning to read words. The 26 musicians in the Sluming study began their musical training at an average of 9.6 years. At that age reading music would have been one of the first things that they learned. They spent an average of 36 hours per week playing their instruments.

The brains of these 26 musicians were compared to those of a group of 26 nonmusicians of comparable age using mMR. The musicians' brains were found to have an enlargement of the cerebral cortex in one specific region. On the basis of the previous paragraph, can you guess what it is? (This is the kind of guessing game that neuroscientists play all the time when they design neuroimaging studies. Sometimes they guess correctly, but sometimes they are surprised.) Compared to the controls, the musicians had 699 cubic millimeters more gray matter in Broca's area, a crucial part of the language system in the left hemisphere. Broca's area is the part that we use to generate fluent grammatical speech. Trained musicians apparently employ this "language region" as a "music region" as well. (This has been confirmed in other fMR studies of musicians.) In particular, as they read and play using printed music, the orchestra musicians must perform a brain activity that in some sense requires them to silently speak music in their brains. Doing this repeatedly, for 36 hours per week on average and over a symphony-performance lifetime that averaged 20.6 years, apparently led to an expansion of synaptic connections that permitted them to perform skillfully, which led to an enlargement of the specific cortical region that they exercised the most. (An mMR study does not permit measurement of synaptic growth. That is only an inference.)

Reading music is a visual/spatial skill, in that it requires the interpretation of musical notes laid out in a precise spatial orientation. Orchestral playing (unlike solo playing of an instrument) also challenges visual/spatial skills in another way. In the layout of a typical orchestra, the first violins and the cellists are arrayed across the front, close to the conductor, with the first violinist in the second-in-command position. Behind them and beside them are the basses, horns, woodwinds, and percussion instruments. As the layout suggests, the musicians must also have a highly developed spatial sense. The first chair violist must (simultaneously) be listening to the left to hear the violins, front and across for the cellos and basses, and behind for the horns, woodwinds, and percussion instruments, while also reading the music and watching the conductor. We know that the parietal lobe, and especially the right parietal lobe, is used for this type of visual/spatial skill. It is somewhat surprising that Sluming's mMR study found no enlargement in that area.

However, a superior ability in that domain was detected in a different way. The subjects in Sluming's study also were given a cognitive test

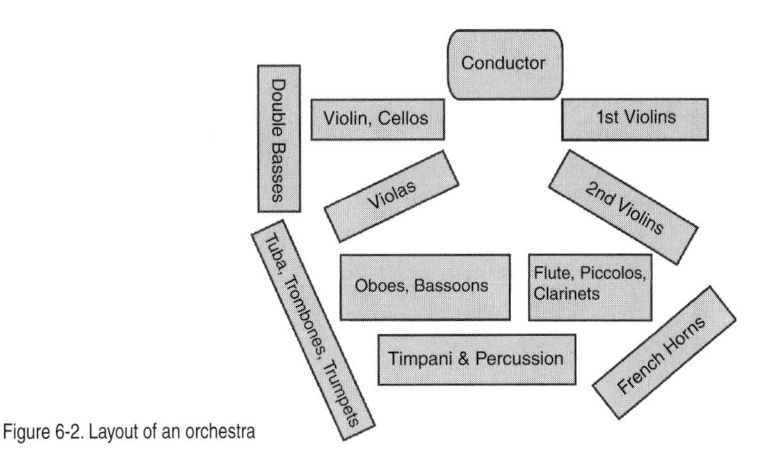

Figure 6-2. Layout of an orchestra

that assesses visual/spatial ability and that is considered to be a classic test of right parietal lobe function. This test, the Judgment of Line Orientation, requires people to estimate subtle differences in the angle between two lines. Thirty different line orientations are used. Of the 26 musicians, 17 achieved perfect scores of 30, whereas only three non-musicians could perform at that level. Thus the orchestra musicians do appear to have enhanced their visual/spatial cognitive skills through practice and performance.

The observation that musical practice and performance may produce structural differences in the brain has been confirmed in other studies as well. In fact, more than 30 neuroimaging studies of musicians of various kinds had been conducted at the time I wrote this book—too many to summarize without turning this chapter into a scholarly "overview article" that would produce a big yawn in most readers. Here are just a few highlights. Although the Sluming study did not show enlargement of the parietal lobe, other studies have indicated that this region is larger or more metabolically active in musicians. Multiple studies have shown that musical performers also have an enlargement of the cerebellum, a part of the brain that is highly developed to monitor motor coordination—and is also now recognized as a monitor of rapid online cognitive activities. Multiple studies have shown that musicians use components of the left hemisphere language system when they read and perform music, and also when they listen to music.

Suffice it to say, these multiple studies contain enough cumulative evidence that we can conclude that studying and performing music are good for the brain. (Those orchestra musicians also had more gray matter overall than did the nonmusicians, and their brains were also less susceptible to gray matter loss with age than were those of the nonmusicians.) Furthermore, these studies indicate that musical practice and performance produce measurable changes in brain structure and function. They support the fact that the brain is plastic, and that creative activities produce specific changes in the brain and in performance on cognitive tests. When we combine these with findings from the taxi driver study, we can also conclude that it may never be too late to start training the brain. While most musicians in these studies began their training at an early age, the taxi drivers didn't start pumping up the synapses in the hippocampus until they were in their twenties.

At this point in the study of creativity and the brain, musical creativity has received by far the greatest emphasis. Although neuroscientists have begun to explore other areas, especially creativity in the visual arts, most of this work has not yet examined the brain directly. The emphasis has primarily been on neuroaesthetics, the study of how we perceive art or how and why artists are drawn to particular shapes or styles, such as Mondrian's preference for straight lines, or Picasso's preoccupation with cubism in an effort to abstract the components of an image or shape. Almost nothing has yet been done to study literary creativity and the brain. (This author has given the topic a lot of thought, but is waiting for inspiration to reveal the perfect experimental design to her.) And studies of the brain and creativity in the sciences are essentially nonexistent.

In short, we know quite a bit, but there is MUCH more yet to be done.

Ordinary Creativity and Extraordinary Creativity

This book is primarily about extraordinary creativity. I wanted to write about how extremely gifted people have created things that have made our lives, our society, and our civilization richer and more beautiful. We have learned that highly creative people have particular personality and cognitive traits, such as openness to experience, curiosity, and a tolerance of ambiguity. We have learned that they often get their ideas as flashes of insight, through moments of inspiration, or by going into a state at the edge of chaos, where ideas float, soar, collide, and connect. We have learned that this creative state arises from a mind and brain that are rich in associative links that encourage new combinations to occur freely. And we have learned that the brain is plastic—that we can change, and hopefully improve, our brains by exercising them.

We have also learned that all of us possess, at a minimum, something I have called ordinary creativity. To call it ordinary is not to diminish it. The fact that we can all generate novel speech "on the fly" is a testament to the "extraordinary ordinary creativity" of our glorious human brains. But many of us manifest creativity in our daily lives in other ways. A teacher who plans novel and interesting ways to excite her students about the beauty of math, or of the laws of physics, or of the poetry of John Donne, is making a significant creative contribution. And if she does it well, their appreciation of the joy of learning will be a gift to them throughout their lives. A businessman who conceives of new ways to interest people in the product he is selling is contributing creatively to his company, and ultimately he may be enriching the local or the world economy. A person who sings in a choir or who plays in an informal rock band or jazz group is doing something creative. So is the mother who makes her children's clothing. Many people who enjoy cooking are creative, envisioning new ways to combine ingredients to produce a tantalizing mixture of flavors or a succulent sensation and texture in the mouth and on the palate. Participating in a book club or reading group, meeting regularly to discuss reactions to a book's ideas and content, is a creative activity. Figuring out how to teach a young son or daughter new ways to learn to manipulate shapes or form abstractions, such as the concept of color, is a creative activity. Ordinary creativity is all around us. We all participate in it in some way.

But wouldn't you like to take what you already have, and figure out how to make yourself even more creative? And, if you are a parent, wouldn't you also like to help your child's young brain grow in ways that will enhance creativity? Here are some ideas that may help to build better brains.

Mental Exercises for Adults

Many of us have exercise programs for our bodies. Very few of us have exercise programs for our minds and brains. If we pause to ponder this fact, it is quite astonishing. Why have we chosen to discipline and develop the body and not the brain? Perhaps it is a reflection of problematic priorities in our contemporary culture. Or perhaps we just haven't been thinking creatively enough.

As I was close to completing this book, I ran across a book chapter by a couple, Robert Root-Bernstein and his wife, Michele Root-Bernstein, who discuss creativity in a refreshingly novel way. They argue that creativity may often arise from taking ideas from one field and transferring them to another, and that mastering one field is a good foundation for mastering another. They also discuss the phenomenon of being a polymath. Like so many good words, "polymath" derives from Greek (*poly* = many, *mathei* = to know, learn). So a polymath is a person who knows many things and who has mastered many fields. They illustrate their argument with examples of creativity from polymaths. (We have seen lots of examples in this book as well.) Ever one to track things down, I discovered that they have also written a recent book about creativity, *Sparks of Genius* (1999), which describes thirteen ways to enhance creativity. Some of those ways were already in my outline for this section of this book, and I have still included them, without guilt, because I discuss them differently and because this section of my book is quite brief. For a longer discussion of the topic, however, I heartily recommend *Sparks of Genius.*

The essence of my recommendations for mental exercises to enhance your own creativity is to set aside some time in your daily life that is devoted exclusively to learning to think and perceive in novel ways. Develop your own "creativity workout," just as you might develop a physical workout. When you first begin to do this, you may find it as difficult as it is for many people to undertake a serious physical exercise program. First, most of us have very little free time. Second, when we do have it, we often feel too tired or drained to take on mental challenges. It is much easier to turn on the TV, read pulp fiction whose content we will barely remember, or surf the Web.

So if you add mental exercises to your daily routine, at first it is likely to feel like hard work, and you will be tempted to give up in fatigue or frustration. As you persist and build up the "mental muscles" in your brain, the exercises will gradually become easier and will even be fun. You may find you get a "thinker's high" that is comparable to a runner's high. And just imagine that gray matter slowly expanding in your brain as you increase those synapses. To further motivate yourself, be aware that neuroscience has already shown that one of the most powerful factors that protects against degenerative brain diseases such as Alzheimer's is higher levels of education. You probably cannot change your years of education at this point, but mental exercise and self-education are the best possible proxies for it.

Here are some suggested exercises. Try to allocate at least thirty minutes per day to doing them. You can probably devote additional time on weekends. Select a combination that you find appealing, but try to do at least two. As you get into the spirit of mental exercise, you may think up some new ones on your own.

Choose a New and Unfamiliar Area of Knowledge and Explore It in Depth

One of the best ways to get a new perspective on things—an important resource for thinking creatively—is to tackle a new field that you know little or nothing about. If your college major was biology or physics, try studying poetry or painting. If you spend your life thinking about computer architecture, try studying history or reading biographies. If you spend your life in the business world, try learning about geography, earth science, or oceanography. If you have longed to learn to play the piano and have had no training in music, start doing it. Just choose something that is quite different from your current interests or occupation.

In this book we have already encountered examples of people who probably owed part of their creativity to their ability to bridge across disciplines that are normally considered disparate. Anatomy is considered a science, and painting and sculpture are considered to be arts. Yet both Michelangelo and Leonardo lived comfortably in both disciplines, and their knowledge of the one enriched the other. Could Michelangelo have created David's beautiful body if he had not spent those secret hours dissecting the human body at San Spirito? What would the Sistine ceiling look like if its painter had been unable to build up the human body by imagining bone, covered by muscles, and in turn covered by skin? If Leonardo or Michelangelo had been born in the twentieth century, what might they have done? Worked on the Manhattan Project? Designed Lincoln Center or the Guggenheim Museum in New York or Bilbao? Discovered the theory of relativity or quantum mechanics? Written novels or philosophical treatises? Done research in neuroscience in order to study the human brain?! Each seems to have been innately endowed with multiple interests and an inability to see boundaries between disciplines that some might consider distant from one another. Very likely they would take the same approach if they lived today.

Mother Nature does not give everyone the gift of being a polymath. But that can be at least partially achieved by nurture. The first step is to learn a new and totally different field. Many people have a secret longing to do something different from the work that is their daily bread. If you have always wanted to try your hand at painting, to take up the violin, or to master a foreign language, take the time and make the effort to do it. Very importantly, don't dabble. Do it in depth and with a passion, for this is the only way that you will really exercise your brain. Many successful people have benefited from working in one field by day and in another in their spare time. Churchill and Eisenhower painted. William Carlos Williams, a doctor, wrote poetry, as did Wallace Stevens, a lawyer. Einstein played the violin. Benjamin Franklin was an inventor, a writer, and a statesman.

Spend Some Time Each Day Practicing Meditation or "Just Thinking"

The creative personality is characterized by openness to experience. How can you practice opening your mind, so that you can intensify your ability to experience life?

One approach is to practice meditation. I personally am not a meditator, although I have several friends who are. From them I know that they find it a useful resource for thinking more creatively. It is also helpful for achieving peace of mind, for improving self-discipline, and for feeling at harmony with the universe. For some, it is a way of moving to a higher level of consciousness, perhaps for getting to "that place" described as part of the creative process in chapter 2. If you decide to take up this approach, many resources are available, ranging from books to tapes to Internet sites. To get started, choose your favorite venue and find an approach that appeals to you.

If you think that meditation is a silly practice that is performed primarily by warmed-over hippies, think again. The study of the effects of meditation on the brain has become a serious area of research in neuroscience, and it indicates that practicing meditation has measurable beneficial effects on brain function. Recent studies have focused on a brain characteristic known as gamma synchrony. Gamma waves are very high frequency oscillations that occur in the brain. When they occur in synchrony in different brain regions, this pattern is thought to reflect the communication of neuronal groups (known as neuronal assemblies) that are widely distributed through the brain and that are engaged in integrating complex information in order to discover its meaning or to solve a problem. Gamma synchrony can be measured by simple electroencephalographic (EEG) techniques.

A research group based in Wisconsin, led by psychologist Richard Davidson, has studied gamma synchrony in Buddhist monks both during their "resting baseline" and while they practiced meditation. The particular kind of meditation studied was "nonreferential" and objectless, in that the monks were not focusing on a particular object or thing. (Some other frequently practiced forms of meditation emphasize concentrating only on one's breathing, or on a mantra, or on a fixed object.) These monks instead practiced a form of meditation that is used in Tibetan Buddhism, which seeks to achieve a state of unconditional loving kindness and compassion that is described as an "unrestricted readiness and availability to help all living beings." During this state of "pure compassion," benevolence and compassion pervade the mind as a way of being. As a comparison group, college students were given one week of training, during which they were taught the techniques of "pure nonreferential compassion" meditation and practiced it for one hour each day.

When the EEGs of the two groups were compared, the monks had markedly higher levels of gamma synchrony—in fact, the highest that have been measured. Apparently their meditative practices also improved the harmony of their brain function when they were not meditating, because they had higher gamma synchrony then as well. Levels of gamma synchrony during baseline and during meditation were both correlated with the number of hours that each monk had spent practicing meditation, suggesting that the degree of training is related to the improvement in "gamma power." And where was gamma power the greatest? You by now can easily guess. It was in the association cortices that are the reservoir of creativity—frontal, temporal, and parietal association regions. The conclusion from Davidson's study is clear and compelling. People can change their brains by training them in the practice of meditation, so that they improve the quality of their moment-to-moment awareness not only during meditation but also during the routine of everyday life.

An alternative way to "open the brain" is quite different from meditation, which emphasizes achieving an intentional focus on an object (in referential meditation) or on a state (in nonreferential meditation). This alternative is "just thinking" in a free and uncensored way.

(This happens to be my favorite way of achieving a more creative state of mind.) This is the state I call "random episodic silent thought," or REST, which I described earlier. If you decide to practice "just thinking," you can do it in many kinds of places, as long as you can dissociate yourself from any kind of input from the outside world. You can do it while lying in bed, reclining in a bathtub, or swimming, for example. The key thing is to let your mind wander freely and to go to "that place" where ideas and images rise to the surface from unconscious or preconscious sources, form or float or fly, ultimately colliding and connecting to create novel associations that cannot occur easily through consciously willed "brute force" cogitation.

Practice Observing and Describing

Every day you probably walk past something that is intrinsically interesting, but that you in fact barely notice. Improving the ability to observe and describe the world around you is another resource for building a better and more creative brain. There are many ways to enjoy variations on this basic theme.

For example, you might specialize in observing a particular sort of thing that you yourself choose. The topic will vary, depending on where you live. It might be birds or barns. It might be paintings in a nearby museum. It might be people you encounter when riding a subway, or the people at the checkout counter in a store where you shop. It might be the way shop windows are organized. It might be the colors and patterns of Oriental rugs. It might be the various flavors that are tasted in red wine. Any of these might be a theme that you follow for a month or more, which is succeeded by a new one after that.

The first stage in this mental exercise is observing. Your long-term goal is to look intently and in detail, so that you observe aspects that you would not normally notice. You might begin, however, with a *gestalt* approach, much as the impressionist painters did when they invented impressionism in nineteenth-century Paris. This will exercise the emotional and intuitive components of your brain. What thoughts and feelings does that building (bird . . . person . . . rug . . . painting) evoke in you? Think, and carefully choose the right words to describe them. Then switch over to observing more analytically. You will want to break your observation down into the details that are appropriate to what you are observing. If it is a person, for example, look at the face . . . the eyes, eyebrows, forehead, nose, lips, skin texture and color; then move on to the hair, the general body appearance and posture, the clothing, the shoes. Choose the perfect words to describe the shapes, colors, and patterns that you are seeing.

While performing this part of the exercise, you are beginning the process of describing while you are observing. But to make the exercise complete you need to produce a written description. Sit down at your computer or at your desk and write a paragraph or two in which you describe your observations in the most elegant and precise language you are able to achieve. (You probably haven't done this sort of thing since you took freshman composition in high school or college.) You may want to buy a thesaurus to consult, so that you can find better words than those that come immediately to mind.

As you continue this exercise over weeks and months, you should work on one topic several times a week for at least one month. As you build up a set of descriptions, notice how you are getting better and better at observing and describing. Then switch to another topic, preferably one that is quite different from the first.

You will very likely begin to notice a payoff from this exercise in your daily life. Your practice will generalize to many other things that you do. Observing carefully, in both an intuitive and an analytic manner, will begin to become habitual. You will be better at sizing up new people whom you encounter, if you do the "observe people" exercise, for example. You will probably be better at remembering faces. Your vocabulary will grow. Your writing will improve. And you will be growing new synapses in your visual, language, and association cortices as a consequence of all this mental exercise.

Practice Imagining

The human brain's capacity for imagining is another resource for creativity.

Consider this paradox.

Objectively, we can only be in one place at one time. I can only be Nancy Andreasen, sitting in my study at my computer, on an evening early in March 2005.

But subjectively I can be anyone, anything, anywhere, and in any place. I can travel to Tudor England and discuss Utopias with Thomas More. I can travel to Kenya and study the social behavior of lions. I can join Lorenzo, Michelangelo, and the members of the Florentine Academy to discuss Neoplatonism in the sculpture garden. I can be the nucleus of the first cell created when a human being is conceived, formulating the instructions for growth that will ensue. I can become even smaller. I can be DNA inside a neuron, sensing that my cell is being repeatedly stimulated, and deciding to express one of my genes that will send protein messengers out to build synapses and create new connections so that my human "owner" can remember. I can be an unmanned space capsule, hurtling through the universe, sensing and observing the sights and sounds that rush past me. I can ride on a photon traveling at the speed of light, as Einstein did, when he developed the theory of relativity. I can be a raccoon washing my food, while watching for larger animals who might perceive me as a tasty treat. I can be a soaring hawk or bald eagle looking for prey. I can be an early hominid, looking for predators while foraging for food, wondering how I can find better tools to survive. I can see a large misshapen block of marble and also see David inside—at least now that I know what he looks like. I can do all this, and much more, by simply exercising the imagination that resides in my brain, while sitting in my study at my computer, on an evening early in March 2005.

Our ability to use our brains to get "outside" our relatively limited personal perspectives and circumstances, and to see something other than the "objective" world, is a powerful gift. Many people fail to realize that they even have this gift, and most who do rarely use it.

If you haven't done it before, spend some time imagining. You'll be surprised at how fun and interesting it is. And even if you have, practice making a practice of it occasionally.

Your opportunities to imagine are much greater than you might imagine. Begin by choosing something that you find inherently interesting. If you like cars, think about what your world would feel like if you viewed it from the perspective of a car's various parts—carburetor, pistons, brakes, steering system, or tires. Imagine how you would feel if your engine were turned on by an insensitive driver who revved the engine too high while your oil was still cold. Then imagine how you would feel if your driver loved you and pampered you like a high-spirited filly. On a human note, imagine what that homeless person, standing on a street corner and begging, is actually experiencing. How did he happen to arrive at this miserable situation? Where did he grow up? What were his parents like? What is he feeling right now? You might approach this scenario from multiple perspectives—one cynical, one forgiving, and one neutral. There are almost endless "imagining exercises" that you can do. You can move through various kinds of objects or living things—cars, antique tables, violins, roses, people from all walks of life—through limitless possibilities. As in the observing exercise, you should write down the products of your imagination on a regular basis.

The essence of the imagining exercise is to expand your perspective on the world so that you are liberated from time and space.

Tips for Teaching Tots

Every parent holding a newborn infant feels nearly overwhelmed by an awesome sense of responsibility for shaping its young life. Given what we now know about the relative importance of nurture in influencing brain plasticity, the responsibility seems even more weighty. Watching that child grow from a helpless newborn to a cooing infant to a crawling explorer to a talkative toddler, and on and on, lays out a fascinating panorama of the processes of brain development, which continues on into early adulthood. Some of this development is hardwired, but much is shaped by experience. Parents can do a great deal to ensure that their child's brain is enriched through the environment that they provide.

As a psychiatrist and neuroscientist (and a parent and even grandparent), I have a few insights about ways to create an optimal environment that will help your child realize his or her maximal creative potential. So, here are some tips.

Turn Off the TV

We examined the impact of graphic visual imagery on young children earlier in this chapter. A concern about the negative effects of watching TV, or too much TV, extends beyond that single issue. In many homes the TV has become a convenient babysitter. Children sit in front of it, mesmerized by its sounds and images—and therefore are not likely to "get into trouble." There are several problems with this scenario.

First, being active and exploratory (getting into trouble) is how a child learns about the world. It is the most natural thing for the child to do. A child explores because her brain is directing her to pick up objects and manipulate them, to examine the spatial relationship created by the pots and pans inside the kitchen cabinets, or to figure out how to stack and unstack canned goods, to discover the contents of wastepaper baskets, or to examine the relative textures of toilet paper and towels. And how fascinating it is to watch that roll of toilet paper unroll and fill the bathroom! All of this can be fairly annoying to the parents who have to do the mopping up operations. But console yourself. These behaviors don't last very long, and they are helping the brain of little Claire or little Owen build concepts such as space, weight, shape, and even gravitational and other mechanical forces. If you are concerned that the exploratory behavior might be dangerous, simply childproof the house by moving anything that is potentially dangerous to levels that your child is unable to reach. And accept the fact that the house may be a bit of a mess for a few years.

Second, sitting in front of a TV is a passive and sedentary activity. Essentially it trains a child's brain to receive, but not to interact. And it trains the body to sit inactively for long periods of time, decreasing the time available for exercising muscles and learning how to coordinate eye-hand movements and large-muscle movements. The United States is currently experiencing an epidemic of childhood obesity. A diet of "junk food" and "empty calories" is one contributor. But the sedentary life of childhood TV addicts is another.

But what about the educational value of TV? Good programs and videos can expose children to places and things that they might never otherwise see, and these media may be a compelling way of introducing new concepts to children. For example, there are a variety of educational videotapes, such as those in the Baby Einstein series. These emphasize learning about music, as in Baby Mozart or Baby Bach. They teach colors, as in Baby Van Gogh. In small doses, these may be good for a child. Just don't overdo it.

In the birth to age five range, give your child minimal exposure to TV or none at all. And keep it minimal after that as well. If the child is exposed, limit the exposure to educational videotapes. A good rule of thumb is no more than an hour per day.

Read Together, Interactively

Recommending reading instead of TV for young children may sound old-fashioned, given that TV is here to stay. But reading is here to stay as well. Our ways of absorbing the written word may gradually change over time and across generations, but everyone will need to be able to read efficiently and with a high degree of comprehension for the foreseeable future. Your child will certainly be doing a great deal of reading in front of a computer terminal, but he or she will also be doing a great deal of learning by reading books as well. Being a skillful reader is a powerful asset for anyone.

Start reading to your child as early as possible. This can begin as early as five or six months. There are lots of good books for kids. Here are some examples (already road-tested by your author on her children or grandchildren or both).

Good Night Moon, by Margaret Wise Brown (6 months–2 or 3 years)
Maisy's Morning on the Farm or *Maisy's Train,* by Lucy Cousins (6 months and up)
Dinosaurs Dinosaurs, by Byron Barton (6 or 12 months–2 years)
Hop on Pop, by Dr. Seuss (12 months–2 or 3 years)
Cars and Trucks and Things that Go, by Richard Scarry (12 months–3 years or more)
Dig Dig Digging, by Margaret Mayo and Alex Ayliffe (12 months and up)
Voyage to the Bunny Planet, by Rosemarie Wells (12 months and up)
Where the Wild Things Are, by Maurice Sendak (12 months and up)
Olivia, by Ian Falconer (18 months–5 years)
The Munchworks Grand Treasury, by Robert Munch (2 or 2.5 and up)

Some of these authors, such as Dr. Seuss or Richard Scarry, have been very productive, so you can find lots more books by them that your child will probably like. And there are lots and lots of other good children's books. Amazon.com has reading lists that are posted by parents, which can be a useful resource for finding others that have been well tested in the real world.

You should read to your child daily, and ideally two or three times per day. At a minimum, make reading a part of the "going to bed experience," so that the day is always finished by reading several books together. When you read with and to your child, don't make it a passive experience. And don't read in a monotonous style. Get into the story. Ask your child to get into it as well, by asking questions related to the content and the pictures. Read a page, and then say: "Where is the mouse?" "Where is the moon?" "Show me where the cheese is." Children love to participate in the reading experience. As they grow older, use reading as a way of building concept formation, such as number, shape, and size. Ask questions like, "How many firemen are in the picture?" "How many gloves are they wearing?" (This kind of interactive reading can be coupled with teaching simple number concepts, such as learning to count. Or with simple reading skills, such as recognizing letters of the alphabet.) Children also love to retell the story after it has been read to them, and this is a good way for them to begin to practice learning retention, a skill they will need to use throughout their lives.

Another important point. Children learn by example. If they see their parents reading, they will want to be able to read books as well. If a home contains bookshelves, and if they see their parents reading and using the books on those shelves, they will be imbued with a tradition of learning. And don't forget to make frequent trips to the public library, as a family activity. Get everyone a borrower's card.

Emphasize Diversity

You as a parent have a choice about what kinds of books and toys your child is exposed to . . . at least during the first five years or so. (This is especially true for children who are given a minimalistically healthy TV diet, and who therefore are not deluged with commercials for junk food and junk toys.) Therefore, you can ensure that your child is exposed to a good mix of toys that will stimulate the brain in a variety of ways. Toy companies are very tuned-in to the contemporary conscientious parent, and they are manufacturing educational toys that are designed—even in a single toy—to teach diverse concepts such as color, shape, letters, numbers, and melodies. Generally, they are bright, colorful, and appealing to most children. (They also use a lot of batteries.) Their appeal, and educational value, will be enhanced if the parent sits beside the child and plays with the toy as well. Parental involvement is the best antidote to the child's becoming bored with an educational toy.

Much as it is good for adults to learn about new topics in depth, it is also good for your child to learn about multiple and diverse topics as he or she grows a bit older. Your goal should be to help little Owen or Claire become a polymath—a child interested in both music and Lego blocks, in both painting and working puzzles, in both arts and sciences. Your choice of toys for your child will help build diverse interests. If Owen focuses on a particular toy with exuberant interest, fine. Some children can even become quite obsessional in their interests. And to some extent, this is a good thing, because it is the nascent beginning of an ability to probe a topic in great depth. A young child may become preoccupied with learning all about kinds of trucks, or kinds of bugs, or kinds of dinosaurs. But your role as a parent is to help Owen achieve balance in his interests as well. The best approach is to choose a toy or topic that is quite different from the usual and make it the play topic for a morning, afternoon, or evening. This may mean pulling out paints or putting on a CD for a sing-along. A good way to capture the attention of your child is to role model an interest in the balanced activity by pulling it out and exploring it yourself.

What about gender differences in play interests? I have many friends who insist that gender differences are real and obvious, and that they arise spontaneously and independently of parental influences. Much of the time these same people engage in classic gender stereotyping in front of their children, pronouncing the son "a real active little boy" or the daughter "such a sweet little girl." At present the solid and rigorous scientific evidence for true gender differences in the brain is minimal. Although some neuroscientists argue that "women are better at verbal tasks and men are better at visuospatial tasks," and although there is some support for this simple stereotype from cognitive testing, the subjects tested have not been raised in a gender-neutral environment. No matter what happens at home (and there is often quite a bit of gender stereotyping there), schools, the media, and churches surround the growing child with gender stereotypes. These suggest, stated simply, that boys are more rambunctious and better at math and science and that girls are more passive and better in social and verbal interactions.

Given that children cannot be reared in a gender-neutral environment, we will never know how much these imputed differences (if true, and that is debatable) are due to socialization, and resultant effects on brain plasticity. If it is assumed that girls want dolls and books rather than Tinkertoys and trucks, that is what they will probably get . . . at home, at the day care or preschool . . . and equivalently once they start school. As a parent, your best chance to let your son become a writer or painter and your daughter an engineer or mathematician is to give them equal opportunities to do both when they are infants and young children. Both boys and girls should be given an equal, and equally diverse, environment that will help their plastic and adaptive brains develop a polymathic approach to life and the world.

Ask Interesting Questions

Intense curiosity is an important attribute of creative people. Children are naturally curious, so this is a trait that you can easily enhance and build on, by teaching them to ask questions.

When you are with your child—reading, eating, talking, walking, or whatever—encourage him or her to look around and ask questions about how things work, or why they are the way they are. The kinds of questions will of course depend on the age of the child. Here are some examples.

Look at a lamp that contains an electric light bulb, and ask: What happens when we turn the switch to make the light turn on and become bright? This leads in turn to a series of other questions, such as What is an electron? What makes it flow? When you travel on an airplane, ask, How can such a heavy plane fly through the air? Again, this leads to other questions, such as What is a vacuum? There are lots of other simpler questions. Why is grass green? What do rabbits eat? Where do they sleep? Where do they live in the winter, and how do they find food then? Why do stars twinkle? What is the difference between a star and a planet? What are the names of the planets? Where are they located in the sky? Why do they appear to move around? What is the sun? What is the moon? Why do people steal? Why do they lie? Why are they sometimes cruel to one another? Why are people nice to each other? Why do people kill each other?

Of course, your child will ask questions too. Sometimes you will not know the answer. If you don't, don't discourage the question. Make it a joint project to figure out the answer. Perhaps you and the child will go to a bookstore or library and get a book on the topic. (This is a good modeling activity to encourage an interest in books.) Perhaps you will check the Internet. The Internet is a wonderful resource, but you will need to be careful about leading a child to think that it is the only, or even the primary, resource. In addition to suggesting the use of books, suggest the use of direct observation. And for some questions, a discussion back and forth is another good approach.

One of my daughters astounded me one evening when I was tucking her into bed for the night. The final stage in our bedtime ritual was for me to sit beside her and tell her a story about something. The theme we were pursuing at the time was stories from the Bible. (I think that whatever one's religious persuasion, it is important to be familiar with biblical literature and history. Otherwise it can be very difficult to appreciate a great deal of art and literature that was inspired by, and is based on, the Bible. Even nonreligious families will enrich their children's knowledge base by familiarizing them with that body of knowledge at some point.)

She was only around three years old. After I had told her the story of the Creation, we talked for a while about the concept of an omnipotent, omnipresent, omniscient, and benevolent God. That day at her Montessori preschool, a child had described how robbers had broken into his home and stolen the TV and his mother's jewelry. The thieves had not been caught, and the child was very upset, because the home had been violated. So my daughter asked: "Mommy, if God controls the whole world, and if he is good, why does he let people do mean things to each other?" All on her own, at a very young age, she had already formulated a classic philosophical/religious question: The Problem of Evil. Out of the mouths of babes come interesting questions.

Go Outdoors and Look at the Natural World

Our world, especially in the United States, has become steadily more urbanized. What is not urbanized is usually suburbanized. As parents, many of us have lost touch with the natural world, from which humankind has arisen and with which we had an intimate connection until the Industrial Revolution drew us increasingly into towns and cities and mechanized occupations. As I write this, I am reminded of the following lines from Wordsworth, his lament on the loss of connection between humanity and nature that he already perceived to be occurring nearly two hundred years ago.

> The world is too much with us; late and soon,
> Getting and spending, we lay waste our powers:
> Little we see in Nature that is ours;
> We have given our hearts away, a sordid boon!

The situation is considerably worse in the twenty-first century. By many accounts children, adolescents, and even adults are scarcely aware that milk, butter, and ice cream come from cows. They only know that food comes from the nearby supermarket.

Many of the creative thinkers who populate this book, such as Leonardo and Michelangelo, began their early lives observing the natural world around them, recording images of it in their brains, and drawing on those images later in their lives and in their art. Others began their careers as naturalists. Francis Galton was a gentleman-explorer whose books *Tropical South Africa* (1853) and *The Art of Travel* (1855) describe the flora and fauna he observed and sketched. His cousin Charles Darwin, also a gentleman-naturalist, created a scientific revolution based simply on his observations of nature accumulated during a trip to the Galápagos Islands, reported in *The Voyage of the Beagle* (1840). There are many reasons for "getting back to nature." Observing nature has been a resource for great science, even when the connections are not obvious. (James Watson, for example, intended to be an ornithologist, but discovered the structure of DNA instead.)

Observing nature makes us think about how and why things happen as they do—such as some of the "interesting questions" mentioned earlier. And, very importantly, observing nature exposes our brains and minds to nearly infinite beauty: flowers, trees, forests, lakes, rivers, mountains, oceans, seashells, and all creatures great and small. Our urban and suburban children may be missing out on an opportunity to appreciate and understand the value of life on earth—in all its many forms—because they so rarely even see it.

Most readers of this book are not likely to be readily able to take their children on hikes in nearby woods on weekends, or canoeing on a nearby pond or river. But most can find smaller opportunities on a regular basis. Part of the trick, as always, is to decide to allocate the time and to make the effort. Identify the resources that you have nearby that will give your child a chance to experience green grass, plants, and animals, rather than gray urban concrete. It might be a visit to an arboretum, or a natural history museum, or a park. If you live in suburbia, take neighborhood walks and look at the trees and plants and flowers. Help your children learn to identify different kinds of birds. (If you live in suburbia, this can be aided by buying a bird feeder and observing who the consumers are.) Plan vacations that occasionally take you to great natural locations—to the ocean, the mountains, or even the Great Plains. Your goal is to help your children learn to understand what they see, but also to become immersed in the grandeur of the natural world. Inspiration is a resource for creativity, and the grandeur of the natural world is one of the most inspiring things that they can experience.

Get Them Interested in Music

All children with normal mental capacities grow up learning to speak and comprehend whatever language their family speaks. Early in life the normal brain develops to a point at which speaking and understanding occur naturally. But reading and writing must be taught.

The ability to use language is a uniquely human capacity. Love and appreciation for the rhythms and tones of music may also be uniquely human, and they may be related to our capacity for language. Evidence for this, described earlier, is that functional imaging studies are now demonstrating that the processing and production of music appear to draw on portions of the "language network" with which neuroscientists are now so familiar. But children do not almost automatically learn (as they do with language) to produce or perform music without special exposure and training. And we do not yet know how important "critical periods" are for learning to understand, enjoy, and perform music.

Given this uncertainty, there are many reasons why parents should hedge their bets and give their children early exposure to music. We have already learned that orchestra musicians have more gray matter in their brains than nonmusicians. Unlike watching TV, which is passive and sedentary, listening to music can be done while children do other activities, such as playing with puzzles or constructing with Legos. This gives them early experience with multitasking and dual processing. They can also sing along with the music that they are hearing, or they can dance to it, thereby exercising multiple networks in their brains. What should they listen to? A balanced mixture of classical and popular, child-oriented music may be best. Why classical? Because it contains complex musical forms and themes that children may perceive intuitively long before they can understand them analytically.

What about formal music education? I personally am a strong advocate for this, and for beginning it at a relatively young age. The Suzuki music program, which permits children to learn to play when they are as young as two or three, is outstanding. For the youngest it emphasizes strings (usually violin), but piano can also be introduced fairly early. Learning to perform on an instrument teaches many things in addition to music: the discipline of practicing, the joy of accomplishing and progressing, the poise of performing in front of others, and the experience of playing in a group. As a child matures and is able to play in an orchestra (or a band, for some instruments), the child learns to work as part of a team. And the child's brain also acquires those synapse-building skills of reading printed music and perceiving visual/spatial relationships.

The Creating Brain: Quo Vadis?

Knowing what we now do about the nature of creativity and about brain plasticity, where should we go from here? How can we nurture both ordinary creativity and extraordinary creativity? We know we can potentially build better brains. How can we do it? The advice above will help individuals who read this book. But what can we do as a society?

The lessons from both history and contemporary neuroscience are clear.

Creativity arises from the brain. Its essence is the ability to perceive and think in original and novel ways. Its seeds may be planted by nature, but nurture helps it germinate, flower, and grow. Nature cannot be easily changed. But nurture is under our own control. Nature can offer the opportunity for greatness, but nurture can either facilitate its growth or stunt it.

Just as the brain, like a garden, creates an overgrowth of dendrites and spines that must be pruned back, so Mother Nature no doubt creates more geniuses than are ultimately permitted to thrive. As of now, many are ignored, neglected, or actively discouraged. If so, for society this is both a tragedy and an opportunity. There is very likely a great deal of creative capital that is going unused, simply for lack of encouragement. This is a challenge both to our educational systems and to society in general.

I often wonder how many great creative minds have been lost, because there was no nurture to help them grow and flourish. Some of these were no doubt lost because they were not born into a cradle of creativity. But others have been lost for other reasons, some of which are under our control. Being a woman, I think about women in particular—half of human society. I firmly believe that women are as intellectually creative as men, and yet society has as yet produced very few women who are great creative geniuses. Very likely this is due to lack of nurture rather than to lack of nature. If I had been born a century ago, I myself would probably not have been allowed to be a doctor or a scientist or a professor, and I would not be writing this book. Can we really afford to waste so much human talent—the potential for genius in half of humanity? We might ask similar questions about children of both sexes who grow up in socially-deprived environments. These questions pose great challenges to our educational systems and to our social structures. Some of the "fix" may be through affirmative action. But a better fix would be to change our educational and social structures so that affirmative action is no longer necessary, so that equal opportunities (and obligations) to make creative contributions are apparent very early in life.

This more radical fix means that we must, as a society, begin to think in more creative ways about how to enhance both extraordinary creativity and ordinary creativity in our educational systems and in our social structures. Our knowledge about the neuroscience of creativity and brain development is sufficiently mature that we can begin to use it to plan rational changes in how we raise and educate our children, and how adults live their lives as well. Determining the details for this type of change would require commissions, blue ribbon panels, white papers, books, and probably additional investments in education, and perhaps even in continuing education for adults. But some generalities about the direction of change are clear.

We must learn more about critical periods in brain development and use this information in our educational programs, and in our family education if public education fails us. The need to learn foreign languages early is an obvious example. But there is probably much more that we will also learn about critical periods as time passes, and the knowledge from neuroscience must be translated into childrearing and educational practices.

We must use our recognition that the brains of geniuses cannot see boundaries between fields such as art and science to implement changes in teaching practices and curricula. Both the arts and the sciences need to be introduced early, introduced equally, and introduced often. Teachers need to learn new ways to show their students how the "silos" of language arts and biology and mathematics are in fact connected, and connected in very interesting ways. This may require new approaches to the training of teachers, in addition to the training of students.

We know that brain plasticity continues throughout life, and that using the brain is good for it. Education should not end with high school or college, or even graduate school or professional school. More opportunities for adult education could also be developed within group settings, such as in the workplace, churches, retirement centers, apartment complexes, or even health clubs. Individuals can do much more to exercise their brains than they currently do, creating the possibility that they can slow aging processes and perhaps lessen their risks for degenerative brain diseases.

These are just several examples of what we already know and what we might do.

Over the coming years, we shall learn more and more about the creating brain . . . about how it thinks, learns, and spontaneously self-organizes. As this knowledge evolves, it is imperative that we use it to find more ways to nurture the creative nature that we all share.

BIBLIOGRAPHY

Albert, R.S., and M. A. Runco. "The Possible Different Personality Dispositions of Scientists and Neuroscientists." In *Scientific Excellence,* edited by D. N. Jackson and J. P. Rushton, 67–97. London: Sage, 1987.

Alvarez, A. *The Savage God: A Study of Suicide*. New York: Random House, 1972.

Amabile, T. M. *Creativity in Context.* Boulder, CO: Westview Press, 1996.

Andreasen, N.C. "Creativity and Psychiatric Illness." *Psychiatric Annals,* 8, no. 3 (1978): 23–45.

——. "Mania and Creativity." In *Mania: An Evolving Concept,* edited by R. H. Belmaker and H. M. van Praag. New York: SP Medical and Scientific Books, 1981.

——. "Creativity and Mental Illness: Prevalence Rates in Writers and Their First–Degree Relatives." *American Journal of Psychiatry,* 144, no. 10 (1987): 1288–92.

——. "Creativity, Cognitive Style, and Mood Disorder: How Are They Related to One Another?" *Nervure,* (1994): 17–24.

——. "Creativity and Mental Illness: A Conceptual and Historical Overview." In *Depression and the Spiritual in Modern Art: Homage to Miró,* edited by J. J. Schildkraut and A. Otero, 2–14. West Sussex: John Wiley and Sons, 1996.

Andreasen, N.C., and I. D. Glick. "Bipolar Affective Disorder and Creativity: Implications and Clinical Management." *Comprehensive Psychiatry,* 29, no. 3 (1988): 207–17.

Andreasen, N. C., D. S. O'Leary, T. Cizadlo, S. Arndt, K. Rezai, G. L. Watkins, L. L. Ponto, and R. D. Hichwa. "Remembering the Past: Two Facets of Episodic Memory Explored with Positron Emission Tomography." *American Journal of Psychiatry,* 152, no. 11 (1995): 1576–85.

Andreasen, N. J. C. *John Donne: Conservative Revolutionary.* Princeton: Princeton University Press, 1967.

Andreasen, N. J. C., and A. Canter. "The Creative Writer: Psychiatric Symptoms and Family History." *Comprehensive Psychiatry,* 15, no. 2 (1974): 123–31.

——. "Genius and Insanity Revisited: Psychiatric Symptoms and Family History in Creative Writers." *Life History in Psychopathology,* 4 (1975):187–210.

Andreasen, N. J. C., and P. S. Powers. "Overinclusive Thinking in Mania and Schizophrenia." *British Journal of Psychiatry,* 125 (1974): 452–56.

——. "Creativity and Psychosis: An Examination of Conceptual Style." *Archives of General Psychiatry* 32, no. 1 (1975): 70–73.

Ashby, W. R. *Design for a Brain: The Origin of Adaptive Behavior.* New York: John Wiley & Sons, 1952.

Bailey, C.H., and E. R. Kandel. "Structural Changes Accompanying Memory Storage." *Annual Review of Physiology,* 55 (1993).

Barron, F., A. Montuori, and A. Barron. *Creators on Creating: Awakening and Cultivating the Imaginative Mind.* New York: Jeremy P. Tarcher/Penguin, 1997.

———. *Creativity and Psychological Health: Origins of Personal Vitality and Creative Freedom.* Princeton: Van Nostrand, 1963.

———. *Creative Person and Creative Process.* New York: Holt, 1969.

———. *Artists in the Making.* New York: Seminar Press, 1972.

Blake, William. *The Poetry and Prose of William Blake.* London: Nonesuch Library, 1956.

Bleuler, E. *Dementia Praecox, or The Group of Schizophrenias,* Translated by J. Zinkin. New York: International Universities Press, 1950.

Boden, M. A., ed. *Dimensions of Creativity.* Cambridge: MIT Press, 1994.

Bohm, D., and L. Nichol, eds. *On Creativity.* London and New York: Routledge, 1998.

Bohm, D., and F. D. Peat. *Science, Order, and Creativity.* 2d ed. London and New York: Routledge, 2000.

Bulmer, M. *Francis Galton: Pioneer of Heredity and Biometry.* Baltimore: The Johns Hopkins University Press, 2003.

Cox, C. M. *Genetic Studies of Genius.* Stanford: Stanford University Press, 1926.

Coleridge, Samuel Taylor. *In Anthology of Romanticism,* ed. Ernest Bernbaum. New York: Ronald Press, 1948.

Crick, F., and C. A. and Koch. "Framework for Consciousness." *Nature Neuroscience,* 6 (2003): 119–26.

Csikszentmihalyi, M. *Creativity: Flow and the Psychology of Discovery and Invention.* New York: HarperCollins, 1996.

Dickinson, Emily. *The Complete Poems of Emily Dickinson.* New York: Little, Brown, and Co., 1960.

Drevdahl, J. E., and R. B. Cattell. "Personality and Creativity in Artists and Writers." *Journal of Clinical Psychology,* 14 (1958): 107–11.

Edelman, G. M. *The Remembered Present: A Biological Theory of Consciousness.* New York: Basic Books, 1989.

Ellis, H. A. *A Study of British Genius.* New York: Houghton–Mifflin, 1926.

Frost, Robert. *The Poetry of Robert Frost.* New York: Henry Holt and Co., 1979.

Galton, F. *Hereditary Genius.* London: Walter Scott, 1892.

Gardner, H. *Art, Mind, and Brain: A Cognitive Approach to Creativity.* New York: Basic Books, 1982.

———. *Frames of Mind: The Theory of Multiple Intelligences.* New York: Basic Books, 1983.

———. ed. *Creating Minds: An Anatomy of Creativity.* New York: Basic Books, 1993.

———. *Intelligence Reframed: Multiple Intelligences for the Twenty–First Century.* New York: Basic Books, 1999.

Gaser, C., and G. Schlaug. "Brain Structures Differ between Musicians and Non–Musicians." *Journal of Neuroscience,* 8:23, no. 27 (2003): 9240–45.

Ghiselin, B., ed. *The Creative Process.* Berkeley and Los Angeles: University of California Press, 1985.

Gleick, J. *Chaos: Making a New Science.* New York: Penguin, 1987.

Goodman, Corey. *See* Tessler–Lavigne and Goodman.

Guilford, J. P. "Creativity." *American Psychologist,* 5 (1950).

Hamilton, I. *Robert Lowell: A Biography.* New York: Random House, 1983.

Henry, G. M., H. Weingartner, and D. L. Murphy. "Idiosyncratic Patterns of Learning and Word Association during Mania." *American Journal of Psychiatry,* 128, no. 5 (1971): 819–25.

Heston, L. "Psychiatric Disorders in Foster Home Reared Children of Schizophrenic Mothers." *British Journal of Psychiatry,* 112 (1966).

———. "The Genetics of Schizophrenic and Schizoid Disease." *Science,* 167, no. 916 (1970): 249–56.

Hirsch, W. *Genius and Degeneration.* New York: D. Appleton, 1896.

Hoffman, D. D. *Visual Intelligence: How We Create What We See.* New York: W. W. Norton, 1998.

Holmes, E. *The Life of Mozart, Including His Correspondence.* London: Chapman and Hall, 1878.

Hutchinson, S., L. H. Lee, N. Gaab, and G. Schlaug. "Cerebellar Volume of Musicians." *Cerebral Cortex,* 13, no. 9 (2003): 943–49.

Hyslop, T. B . *The Great Abnormals.* New York: George H. Doran, 1925.

Jamison, K. R. "Mood Disorders and Patterns of Creativity in British Writers and Artists." *Psychiatry,* 52, no. 2 (1989): 125–34.

Janata, P., and S. T. Grafton. "Swinging in the Brain: Shared Neural Substrates for Behaviors Related to Sequencing and Music." *Nature Neuroscience,* 6, no. 7 (2003): 682–87.

Janata, P., J. L. Birk, J. D. Van Horn, M. Leman, B. Tillmann, and J. J. Bharucha. "The Cortical Topography of Tonal Structures Underlying Western Music." *Science,* 298, no. 5601 (2002): 2167–70.

Juda, A. "The Relationship between Highest Mental Capacity and Psychic Abnormalities." *American Journal of Psychiatry,* 106 (1949).

———. *Hoechstbegabung: Ihre Erbverhaeltnisse sowie ihre Beziehungen zu Psychischen Anomalien.* Munich: Urban und Schwarzenberg, 1953.

Judd, L. L., B. Hubbard, D. S. Janowsky, L. Y. Huey, and K. I. Takahashi. "The Effect of Lithium Carbonate on the Cognitive Functions of Normal Subjects." *Archives of General Psychiatry,* 34, no. 3 (1971): 355–57.

Kagan, J., ed. *Creativity and Learning.* Boston: Beacon Press, 1967.

Kandel, E. R. *See* Bailey and Kandel.

Karlsson, J. L. "Genetic Association of Giftedness and Creativity with Schizophrenia." *Hereditas,* 66, no.2 (1970): 177–82.

Roberts, Royston M. *Serendipity, Actual Discoveries in Science.* New York: John Wiley and Sons, 1984.

Killian, G. A., P. S. Holzman, J. M. Davis, and R. Gibbons. "Effects of Psychotropic Medication on Selected Cognitive and Perceptual Measures." *Journal of Abnormal Psychology,* 93, no. 1 (1984): 58–70.

Koestler, A. *The Act of Creation.* New York: Macmillan, 1964.

Lange, W., E. Paul, and C. Paul.*The Problem of Genius.* London, K. Paul, Trench, Trubner, 1931.

Lind, R. D. *Lyric Poetry of the Italian Renaissance.* New Haven: Yale University Press, 1954.

Lombroso, C. *The Man of Genius.* London: W. Scott; New York: Charles Scribner's Sons, 1891.

Lowell, R. *Life Studies, and For the Union Dead.* New York, Noonday Press, 1967.

Lowes, John Livingston. *The Road to Xanadu: A Study in the Ways of Imagination.* Boston and New York: Houghton–Mifflin, [1927] 1964.

Lutz, A., L.L. Greishar, N.B. Rawlings, M. Ricard, and R.J. Davidson. "Long–term meditators self–induce high–amplitude gamma synchrony during mental practice." *PNAS*, 101, no. 46 (2004): 16360–73.

Mackinnon, D. W. "Personality and the Realization of Creative Potential." *American Psychologist,* 20 (1965): 273–81.

Macrosson, W. D. K., and P. E. Stewart. "The Inclination of Artists to Partition Line Sections in the Golden Ratio." *Perceptual and Motor Skills,* 84 (1997): 707–13.

Maguire, E., D. Gadian, I. Johnsrude, C. Good, J. Ashburner, R. Frackowiac, and C. Frith. "Navigational Related Structural Change in the Hippocampi of Taxi Drivers." *Proceedings of the National Academy of Sciences,* 97 (2000): 4398–4403.

McNeil, T. F. "Prebirth and Postbirth Influence on the Relationship between Creative Ability and Recorded Mental Illness." *Journal of Personality,* 39, no. 3 (1971): 391–406.

Merton, R. K., and E. Barber. *The Travels and Adventures of Serendipity.* Princeton: Princeton University Press, 2004.

Meyers, M. A. "Science, Creativity, and Serendipity." *American Journal of Roentgenology,* 165 (1995): 755–64.

Mozart. *See* Holmes.

Nasar, S. *A Beautiful Mind.* New York: Simon and Schuster, 1998.

Newmarch, R. *Life and Letters of Peter Ilyich Tchaikovsky.* London: John Lane, 1906.

Nisbet, J. F. *The Insanity of Genius.* New York: Charles Scribner's Sons, 1900.

Pfenninger, K. H., and V. R., Shubik, eds. *The Origins of Creativity.* Oxford: Oxford University Press, 2001.

Pickering, G. W. *Creative Malady: Illness in the Lives and Minds of Charles Darwin.* London, Allen and Unwin, 1974.

Poincaré, Henri. *The Value of Science: Essential Writings of Henri Poincaré.* New York: Random House, 2001.

Post, F. "Creativity and Psychopathology: A Study of 291 World–Famous Men." *British Journal of Psychiatry,* 165 (1994): 22–34.

Rakic, P. "Specification of Cerebral Cortical Areas." *Science,* 241 (1988).

Richards, R. L., D. K. Kinney, I. Lunde, M. Benet, and A. P. Merzel. "Creativity in Manic–Depressives, Cyclothymes, Their Normal Relatives, and Control Subjects." *Journal of Abnormal Psychology,* 97, no. 3 (1988): 281–88.

Roe, A. "The Personality of Artists." *Educational Psychology Measures,* 6 (1946): 401–8.

———. "A Psychological Study of Physical Scientists." *Genetic Psychology Monographs,* 43 (1951).

Root–Bernstein, R. M. *Sparks of Genius The Thirteen Thinking Tools of the World's Most Creative People.* Boston: Houghton Mifflin, 1999.

Rosenthal, D., and S. Kety, eds. *The Transmission of Schizophrenia: Proceedings of the Second Research Conference of the Foundations' Fund for Research in Psychiatry, Dorado, Puerto Rico, 26 June to 1 July 1967.* Oxford and New York: Pergamon Press, 1968.

Rosenthal, D, P. H. Wender, S. S. Kety, and F. Schulsinger. "The Adopted–Away Offspring of Schizophrenics." *American Journal of Psychiatry,* 128 (1971): 307–11.

Rosner, S., and L. E. Abt. *The Creative Experience.* New York: Grossman, 1970.

Rothenberg, A. "The Process of Janusian Thinking in Creativity." *Archives of General Psychiatry*, 24 (1971)195–205.

——. "Psychopathology and Creative Cognition: A Comparison of Hospitalized Patients, Nobel Laureates, and Controls." *Archives of General Psychiatry*, 40 (1983).

Schildkraut, J.J., Hirshfeld, A.J., Murphy, J.M. "Mind and Mood in Modern Art, II: Depressive Disorders, Spirituality, and Early Deaths in the Abstract Expressionist Artists of the New York School." *American Journal of Psychiatry*, 151 (1994): 482–488.

Schou, M. "Artistic Productivity and Lithium Prophylaxis in Manic–Depressive Illness." *British Journal of Psychiatry*, 135 (1979): 97–103.

Schuldberg, D., C. French, B. L. Stone, and J. Heberle. "Creativity and Schizotypal Traits: Creativity Test Scores and Perceptual Aberration, Magical Ideation, and Impulsive Nonconformity." *Journal of Nervous and Mental Disorders*, 176, no. 11 (1988): 648–57.

Seagoe, M. V. *Terman and the Gifted.* Los Altos, CA: William Kaufmann, 1975.

Shakespeare, W. *The Complete Works of Shakespeare.* Edited by G. L. Kittredge. Boston: Ginn, 1936.

Shaw, E. D., J. J. Mann, P. E. Stokes, and A. Z. Manevitz. "Effects of Lithium Carbonate on Associative Productivity and Idiosyncrasy in Bipolar Outpatients." *American Journal of Psychiatry*, 143, no. 9 (1986): 1166–69.

Simonton, D. K. *Genius, Creativity, and Leadership: Historiometric Inquiries.* Cambridge: Harvard University Press, 1984.

——. *Greatness: Who Makes History and Why.* New York: Guilford Press, 1994.

——. *Origins of Genius: Darwinian Perspectives on Creativity.* Oxford: Oxford University Press, 1999.

——. *Creativity in Science: Chance, Logic, Genius, and Zeitgeist.* Cambridge: Cambridge University Press, 2004.

Sluming, V., T. Barrick, M. Howard, E. Cezayirli, A. Mayes, and N. Roberts. "Voxel–based Morphometry Reveals Increased Gray Matter Density in Broca's Area in Male Symphony Orchestra Musicians." *Neuroimage*, 17, no. 3 (2002): 1613–22.

Snow, C. P. *The Two Cultures.* Cambridge: Cambridge University Press, 1959.

Spender, Stephen. *World WithinWorld: The Autobiography of Stephen Spender.* New York: Modern Library, 2001.

Stern, Richard. *See* Hamilton.

Sternberg, R. J., ed. *The Nature of Creativity.* Cambridge: Cambridge University Press.

——. 1999. *Handbook of Creativity.* Cambridge: Cambridge University Press, 1988.

Sternberg, R. J., E. L. Grigorenko, and J. L. Singer, eds. *Creativity: From Potential to Realization.* Washington, D.C.: American Psychological Association, 2004.

Taylor, C. W., and F. X. Barron. *Scientific Creativity, Its Recognition and Development: Selected Papers from the Proceedings of the First and Second Utah Creativity Research Conference.* New York: John Wiley and Sons, 1963.

——. *Scientific Creativity.* New York: John Wiley and Sons, 1964.

Tchaikovsky. *See* Newmarch.

Terman, L. W. "The Intelligence Quotient of Francis Galton in Childhood." *American Journal of Psychology*, 28 (1917): 209–15.

——. *Genetic Studies of Genius.* 6 vols. Stanford: Stanford University Press, 1925–59.

Tessler–Lavigne, M., and C. S. Goodman. "The Molecular Biology of Axon Guidance." *Science,* 274 (1996): 1123–32.

Vasari, G. *Lives of the Painters, Sculptors and Architects.* Translated by Gaston du C. de Vere. New York: Alfred A. Knopf, [1550] 1996.

Venables, P. H. "Input Dysfunction in Schizophrenia." In *Progress in Experimental Personality Research,* edited by B. A. Maher and W. B. Maher. New York: Academic Press, 1964.

Vernon, P. E. *Creativity: Selected Readings.* Harmondsworth: Penguin, 1970.

Waldrop, M. M. *Complexity: The Emerging Science at the Edge of Order and Chaos.* New York: Simon and Schuster, 1992.

Weiner, N. *The Human Use of Human Beings: Cybernetics and Society.* Boston: Houghton Mifflin, 1954.

Zatorre, R. J., and C. L. Krumhansl. "Neuroscience: Mental Models and Musical Minds." *Science,* 17, no. 3 (2002): 1613–22.

Zeki, S. *Inner Vision: An Exploration of Art and the Brain.* Oxford: Oxford University Press, 1999.

INDEX

OTHER DANA PRESS BOOKS AND PERIODICALS

BOOKS FOR GENERAL READERS

(available in bookstores)

THE ETHICAL BRAIN
Michael S. Gazzaniga, Ph.D.

Explores how the lessons of neuroscience can help us resolve today's ethical dilemmas, ranging from when life begins to "off-label" use of drugs, such as Ritalin, by students preparing for exams, and from free will and personal responsibility to public policy and religious belief. The author, a pioneer in cognitive neuroscience, is a member of the President's Council on Bioethics. 225 pp. $25.00

FATAL SEQUENCE: The Killer Within
Kevin J. Tracey, M.D.

An easily understood account of the body's ability to go into the fatal spiral of sepsis, a crisis that most often affects patients fighting off nonfatal illness or injury. Tracey puts the scientific and medical story of sepsis in the context of his battle to save a burned baby, a sensitive telling that renders cutting-edge science human and unforgettable. 225 pp. $23.95

A GOOD START IN LIFE: Understanding Your Child's Brain and Behavior from Birth to Age Six
Norbert Herschkowitz, M.D., and Elinore Chapman Herschkowitz

Updated with the latest information and new material, the authors show us how young children learn to live together in family and society and explain how brain development shapes a child's personality and behavior, discussing appropriate rule-setting, the child's moral sense, temperament, language, playing, aggression, impulse control, and empathy.

Cloth, 283 pp. $22.95

(Updated version with 13 illustrations) Paper, 312 pp. $13.95

A WELL-TEMPERED MIND: Using Music to Help Children Listen and Learn
Peter Perret and Janet Fox

Five musicians enter elementary school classrooms, helping children learn about music and contributing both to higher enthusiasm and improved academic performance. This charming story gives us a taste of things to come in one of the newest areas of brain research: the effect of music on the brain. 225 pp., 12 illustrations. $22.95

BACK FROM THE BRINK: How Crises Spur Doctors to New Discoveries about the Brain
Edward J. Sylvester

Goes into two academic medical centers, Columbia's New York Presbyterian and Johns Hopkins Medical Institutions, to watch a new breed of doctor, the neurointensivist, save patients with life-threatening brain injuries. 16 illustrations/photos. 296 pp. $25.00

BEYOND THERAPY:
Biotechnology and the Pursuit of Happiness. A *Report of the President's Council on Bioethics*
Special Foreword by Leon R. Kass, M.D., Chairman. Introduction by William Safire

Can biotechnology satisfy deep and familiar human desires—for better children, superior performance, ageless bodies, and happy souls? This landmark report says these possibilities present us with profound ethical challenges and choices. 376 pp. $10.95

IN SEARCH OF THE LOST CORD: Solving the Mystery of Spinal Cord Regeneration
Luba Vikhanski

The story of the scientists and science involved in the international scientific race to find ways to repair the damaged spinal cord and restore movement. 21 photos; 12 illustrations. 269 pp. $27.95

KEEP YOUR BRAIN YOUNG : The Complete Guide to Physical and Emotional Health and Longevity
Guy McKhann, M.D., and Marilyn Albert, Ph.D.

Every aspect of aging and the brain: changes in memory, nutrition, mood, sleep, and sex, as well as the later problems in alcohol use, vision, hearing, movement and balance. Cloth 304 pp. $24.95 / Paper 304 pp. $15.95

STATES OF MIND: New Discoveries about How Our Brains Make Us Who We Are
Roberta Conlan, Editor

Adapted from the Dana/Smithsonian Associates lecture series by eight of the country's top brain scientists, including the 2000 Nobel laureate in medicine, Eric Kandel. Cloth 214 pp. $24.95 / Paper 224 pp. $18.95

STRIKING BACK AT STROKE: A Doctor-Patient Journal
Cleo Hutton, and Louis R. Caplan, M.D.

A personal account with medical guidance for anyone enduring the changes that a stroke can bring to a life, a family, and a sense of self. 15 illustrations. 240 pp. $27.00

THE BARD ON THE BRAIN: Understanding the Mind Through the Art of Shakespeare and the Science of Brain Imaging
Paul Matthews, M.D., and Jeffrey McQuain, Ph.D. Foreword by Diane Ackerman

Explores the beauty and mystery of the human mind and the workings of the brain, following the path the Bard pointed out in 35 of the most famous speeches from his plays. 100 illustrations. 248 pp. $35.00

THE DANA GUIDE TO BRAIN HEALTH
Floyd E. Bloom, M.D., M. Flint Beal, M.D., and David J. Kupfer, M.D., Editors. Foreword by William Safire

A home reference on the brain edited by three leading experts, collaborating with 104 distinguished scientists and medical professionals. In easy to understand language with cross references and advice on 72 conditions such as autism, Alzheimer's disease, multiple sclerosis, depression, and Parkinson's disease. 16 full-color pages and more than 200 black and white illustrations. 768 pp. $45.00

THE END OF STRESS AS WE KNOW IT
Bruce McEwen, Ph.D., with Elizabeth Norton Lasley. Foreword by Robert Sapolsky

How brain and body work under stress and how it is possible to avoid stress's debilitating effects. Cloth 239 pp. $27.95 / Paper 262 pp. $19.95

THE LONGEVITY STRATEGY: How to Live to 100 Using the Brain-Body Connection
David Mahoney and Richard Restak, M.D. Foreword by William Safire

Advice on the brain and aging well. Cloth 250 pp. $22.95 / Paper 272 pp. $14.95

THE SECRET LIFE OF THE BRAIN
Richard Restak, M.D. Foreword by David Grubin

Companion book to the PBS series of the same name, exploring recent discoveries about the brain from infancy through old age. 201 pp. $35.00

UNDERSTANDING DEPRESSION: What We Know and What You Can Do About It
J. Raymond DePaulo Jr., M.D., and Leslie Alan Horvitz. Foreword by Kay Redfield Jamison, Ph.D.

What depression is, who gets it and why, what happens in the brain, troubles that come with the illness, and the treatments that work. Cloth 304 pp. $24.95 / Paper 296 pp. $14.95

NEUROETHICS SERIES

NEUROETHICS: Mapping the Field. Conference Proceedings.
Steven J. Marcus, Editor

Proceedings of the landmark 2002 conference organized by Stanford University and the University of California, San Francisco, at which more than 150 neuroscientists, bioethicists, psychiatrists and psychologists, philosophers, and professors of law and public policy debated the implications of neuroscience research findings for individual and societal decision-making. 50 illustrations. 367 pp. $10.95

NEUROSCIENCE AND THE LAW: Brain, Mind, and the Scales of Justice

Brent Garland, ed.; Foreword by Mark Frankel. With commissioned papers by Michael S. Gazzaniga, Ph.D. and Megan S. Steven; Laurence Tancredi, M.D., J.D.; Henry T. Greely, J.D.; and Stephen J. Morse, J.D., Ph.D.

How discoveries in neuroscience influence criminal and civil justice, based on an invitational meeting of 26 top neuroscientists, legal scholars, attorneys, and state and federal judges convened by the Dana Foundation and the American Association for the Advancement of Science. 226 pp. $8.95

FREE EDUCATIONAL BOOKS (order through www.dana.org)

ACTS OF ACHIEVEMENT : The Role of Performing Arts Centers in Education.

Profiles of 60-plus programs, eight extended case studies, from urban and rural communities across the United States, illustrating different approaches to performing arts-education programs in school settings. Black and white photos throughout. 164 pp.

PLANNING AN ARTS-CENTERED SCHOOL: A Handbook

A practical guide for those interested in creating, maintaining, or upgrading arts-centered schools. Includes curriculum and development, governance, funding, assessment, and community participation. Black and white photos throughout. 164 pp.

THE DANA SOURCEBOOK OF BRAIN SCIENCE: Resources for Secondary and Post-Secondary Teachers and Students.

A basic introduction to brain science, its history, current understanding of the brain, new developments, and future directions. 16 color photos; 29 black and white photos; 26 black and white illustrations. 164 pp.

THE DANA SOURCEBOOK OF IMMUNOLOGY

An introduction to how the immune system protects us, what happens when it breaks down, the diseases that threaten it, and the unique relationship between the immune system and the brain.

PERIODICALS

ARTS EDUCATION IN THE NEWS

A quarterly, eight-page newspaper reprinting news and feature articles about performing arts education in the schools from leading U.S. and foreign newspapers and magazines.

Free: by mail only; order through www.dana.org/books/press

BRAIN CONNECTIONS: Your Source Guide to Information on Brain Disease and Disorders

Pocket-size, 48-page booklet listing more than 275 organizations that help people with brain-related disorders and those responsible for their care and treatment. A publication of the Dana Alliance for Brain Initiatives. Listings include toll-free numbers, Web site and e-mail addresses and regular mailing addresses.

Free: order through or download from www.dana.org/books/press

BRAIN IN THE NEWS

A monthly, eight-page newspaper reprinting news and feature articles about brain research from leading newspapers and magazines in the United States and abroad during the previous month.

Free: by mail only; order through www.dana.org/books/press

BRAINWORK: The Neuroscience Newsletter

A bimonthly, full-color, eight-page newsletter for general readers reporting the latest findings in brain research.

Free: order through or download from www.dana.org/books/press

CEREBRUM: The Dana Forum on Brain Science

A quarterly journal for general readers with feature articles, debates, and book reviews dealing with the latest discoveries about the brain and their implications for individuals and society.

Subscription (4 issues): $30/year ($42 foreign; $48 institutions)

To order a trial copy: call 1-877-860-0901 or write to

Cerebrum Subscriber Services, P.O. Box 573, Oxon Hill, MD 20745-0573

IMMUNOLOGY IN THE NEWS

A quarterly, eight-page newspaper reprinting news and feature articles about immune system research, disease treatment and prevention, and biodefense from U.S. and foreign newspapers and scientific journals.

Free: by mail only; order through www.dana.org/books/press

PROGRESS REPORT ON BRAIN RESEARCH

Published in March annually since 1995, the Progress Report, a publication of the Dana Alliance for Brain Initiatives, identifies the most significant findings in brain research from the previous year.

Free: order through or download from www.dana.org/books/press